Principles and Technology of
Hydrothermal Refinery of Lignocellulosic Biomass

木质纤维素水热炼制原理与技术

石宁　编著

化学工业出版社
·北京·

内 容 简 介

本书依据作者的研究成果及国内外木质纤维素的研究进展，详细介绍了木质纤维素水热炼制原理与技术。具体包括木质纤维素的基本概念、纤维素水热催化转化制备平台化学品的基本原理、两相体系中纤维素转化制备 5-羟甲基糠醛、含氧化合物水热降解生产焦炭的机理，以及葡萄糖和呋喃类物质水热转化生成的可溶性副产物的鉴定。

本书可供从事新能源、化工、生物工程、材料及相关学科的研究人员和技术人员阅读，尤其适用于木质纤维素资源高值化利用领域的科研工作者参考使用。

图书在版编目（CIP）数据

木质纤维素水热炼制原理与技术/石宁编著.—北京：化学工业出版社，2020.8

ISBN 978-7-122-37585-8

Ⅰ.①木… Ⅱ.①石… Ⅲ.①木纤维-纤维素-热降解-水热法-研究 Ⅳ.①TQ352.6

中国版本图书馆 CIP 数据核字（2020）第 155724 号

责任编辑：韩霄翠　仇志刚　　　　　文字编辑：王云霞　陈小滔
责任校对：王素芹　　　　　　　　　装帧设计：王晓宇

出版发行：化学工业出版社（北京市东城区青年湖南街 13 号　邮政编码 100011）
印　　装：北京盛通商印快线网络科技有限公司
710mm×1000mm　1/16　印张 13¼　字数 228 千字　2020 年 11 月北京第 1 版第 1 次印刷

购书咨询：010-64518888　　　　　　售后服务：010-64518899
网　　址：http://www.cip.com.cn
凡购买本书，如有缺损质量问题，本社销售中心负责调换。

定　价：88.00 元　　　　　　　　　　　　　　　版权所有　违者必究

前 言

在现代工业开始之前，生物质资源一直都是人类食物、能源和材料的最主要来源。自工业革命以来，煤炭、石油和天然气等高能量密度且易于开采加工的化石资源的开采和利用，为人类文明数百年来的突飞猛进和今日的繁荣提供了物质和能源保障。然而，化石资源的开采消耗及其利用过程中所带来的环境问题逐渐成为人类社会发展的障碍。一方面，化石资源为不可再生资源，大规模地开采和消耗必然导致资源的短缺，从而无法满足人类社会发展的需求；另一方面，大量化石资源的利用会带来一系列的环境问题，如气候变暖、酸雨、海洋污染、大气污染、固体废物污染等，也成为人类生存的重要威胁。

为了应对过度依赖不可再生的化石资源所带来的资源和环境问题，世界各国都积极投入对可再生资源的研究，并开发出多种对核能、太阳能、水能、生物质能、风能、地热能、海洋能等可再生资源的利用技术。在上述可再生资源中，只有生物质资源能够像化石资源那样通过一系列物理变化和化学变化得到能源和有机化学品。因此，开发利用生物质资源，对人类文明的可持续发展至关重要。

生物质资源通常指来源于生命体的有机物质，包括动物、植物和微生物中的有机物。但是，动物、植物和微生物中的物质组成存在明显的区别，且植物的果实和茎干也存在明显的区别。比如，动物尸体中含有大量的油脂和蛋白质，某些植物的果实中含有大量的油脂，某些植物的果实中则含有大量淀粉，而植物的茎干则通常由纤维素、半纤维素和木质素组成。由于来源不同的生物质的化学成分明显不同，所以其利用方式也存在差异。相比较于其他生物质，以农作物秸秆和林业废弃物为代表的木质纤维素是一类存量巨大、成分单一的生物质。通常而言，植物茎干的主要成分都是纤维素、半纤维素和木质素，其转化过程具有较为明显的规律。因此，国内外学者在木质纤维素资源的利用方面开展了大量的研究。

水热炼制技术是一种木质纤维素转化技术。在水热炼制技术中，木质纤维素中的多糖组分（包括纤维素和半纤维素）发生水解生成葡萄糖、木糖等单糖，而这些单糖则可以在催化剂作用下选择性地转化生成一系列重要的平台化学品，如5-羟甲基糠醛、糠醛、乙酰丙酸、乳酸、乙二醇、1,2-丙二醇、山梨醇、木糖醇等，这些平台化学品可以作为化工原料生产下游的有机化工品及液体燃料。由于这一领域在促进人类社会可持续发展中所展现出的价值，目前国内外对此领域开展了广泛的研究，多种技术路线正在从实验室走向工厂。

  本书是笔者在对木质纤维素水热炼制技术进行深入研究的基础上，结合国内外同行的研究成果而编著。全书共五章，第一章对木质纤维素的概念、结构及其利用技术进行概述；第二章详细讨论了木质纤维素中的纤维素组分水热转化制备平台化学品（5-羟甲基糠醛、乙酰丙酸、乳酸、乙二醇、1,2-丙二醇、山梨醇等）的相关理论基础和研究进展；第三章论述了笔者在催化纤维素制备5-羟甲基糠醛中的研究成果及所遇到的困难；第四章研究了碳水化合物及其衍生物在水热条件下的降解行为并对水热焦炭的结构进行表征，提出碳水化合物在水热转化过程中生成的α-羰基醛是形成水热焦炭的关键中间体，而羟醛缩合反应是生成水热焦炭的关键步骤；第五章论述了利用LC-MS和LC-MS^{2}研究葡萄糖及三种生物质衍生呋喃类物质（5-羟甲基糠醛、糠醛和糠醇）经水热降解生成的可溶性物质的结构，并分析了碳水化合物的水热转化路径和各种鉴定出的可溶性物质的生成机理。

  本书适用于木质纤维素资源高值化利用领域的相关科研工作者。因笔者水平有限，书中不足和欠妥之处在所难免，敬请读者见谅并予以批评指正。

编著者

2020年6月

目录

第一章
001 概述

1.1 生物质和木质纤维素的定义　002

1.2 木质纤维素的种类及其化学结构　002

  1.2.1 木质纤维素类的种类　002

  1.2.2 木质纤维素的化学结构　003

1.3 木质纤维素的利用　012

参考文献　014

第二章
017 纤维素水热催化转化制备平台化学品的基本原理

2.1 引言　018

2.2 纤维素的催化水解反应　019

  2.2.1 纤维素催化水解的反应介质　019

  2.2.2 纤维素催化水解的催化剂　019

2.3 催化碳水化合物转化制备 5-羟甲基糠醛的原理　024

  2.3.1 5-羟甲基糠醛简介　024

  2.3.2 己糖制备 5-羟甲基糠醛的反应机理　025

  2.3.3 反应介质、催化剂及反应原料　026

2.4 木质纤维素制备乙酰丙酸及其衍生酯类的原理　032

  2.4.1 乙酰丙酸简介　032

  2.4.2 己糖生成乙酰丙酸的原理　033

  2.4.3 木质纤维素及其衍生碳水化合物生成乙酰丙酸的催化剂及生产工艺　034

  2.4.4 利用木质纤维素制备乙酰丙酸酯　036

2.5 催化木质纤维素及其衍生碳水化合物转化制备乳酸（酯）的原理 038
　　2.5.1 乳酸简介 038
　　2.5.2 碳水化合物制备乳酸（酯）的反应路径 039
　　2.5.3 反应原料及反应体系 040
　　2.5.4 碳水化合物转化为乳酸（酯）的催化剂 040
　　2.5.5 碳水化合物水热转化生成其他 α-羟基酸 046
2.6 纤维素制备己糖醇（山梨醇和甘露醇） 048
2.7 纤维素制备乙二醇和 1,2-丙二醇 050
　　2.7.1 纤维素制备乙二醇和 1,2-丙二醇的路径 051
　　2.7.2 纤维素制备乙二醇和 1,2-丙二醇的催化体系 052
2.8 碳水化合物在水热转化过程中发生的基本反应 054
2.9 葡萄糖异构化机理 056
　　2.9.1 碱催化葡萄糖异构化为果糖 056
　　2.9.2 均相路易斯酸催化葡萄糖异构化的机理 057
　　2.9.3 非均相路易斯酸 Sn-β 催化葡萄糖异构化的机理 060

参考文献 062

第三章
085　两相体系中转化纤维素制备 5-羟甲基糠醛

3.1 引言 086
3.2 转化纤维素制备 5-羟甲基糠醛的过程及产物的分析表征 087
3.3 在两相体系中金属盐催化剂的筛选 088
3.4 $NaHSO_4$-$ZnSO_4$ 协同催化纤维素降解制备 5-羟甲基糠醛 090
　　3.4.1 反应时间的影响 091
　　3.4.2 反应温度对纤维素转化的影响 094
　　3.4.3 水/有机相的体积比对反应的影响 094
　　3.4.4 反应体系对其他可溶性糖的转化 095
　　3.4.5 原料用量对 5-羟甲基糠醛收率的影响 097
3.5 反应体系分析及液膜催化概念的提出 097
3.6 结论 099

参考文献 099

第四章
102 含氧化合物水热降解生成焦炭的机理

4.1 引言 103

4.2 模型化合物的水热转化过程及产物的分析表征 106
 4.2.1 模型化合物的水热转化过程 106
 4.2.2 原料的定量分析方法 106
 4.2.3 焦炭的结构表征方法 107

4.3 呋喃衍生物的水热降解行为及降解途径 107

4.4 碳水化合物的水热降解行为研究及其
 转化路径分析 113

4.5 碳水化合物的其他衍生物的水热降解行为 116

4.6 $C_2 \sim C_4$ 短链含氧有机物的水热降解行为 119

4.7 乙二醛和丙酮醛的水热降解路径分析 121

4.8 水热焦炭的结构分析 123
 4.8.1 水热焦炭的元素分析 123
 4.8.2 水热焦炭的 FT-IR 分析 125
 4.8.3 水热焦炭的固态 ^{13}C NMR 分析 126

4.9 水热焦炭的形成机理 128

4.10 水热焦炭的分子结构 132

4.11 抑制碳水化合物水热降解过程中
 水热焦炭的形成 133

4.12 结论 133

参考文献 134

第五章
140 葡萄糖及生物质衍生呋喃类物质水热转化生成的可溶性副产物鉴定

5.1 引言 141

5.2 LC-MS^n 对可溶性葡萄糖水热转化产物的分析条件 143

5.3 LC/MS 和 LC/MS^2 谱图的分析过程 143

5.4 葡萄糖水热转化生成的水溶性物质鉴定 147
 5.4.1 LC-MS 确定葡萄糖降解生成的水溶性
  化合物的化学式 147
 5.4.2 根据 MS^2 质谱图确定化合物的分子结构 150

  5.4.3 葡萄糖水热转化过程中可溶性碳环类物质的形成机理 155

 5.5 呋喃类化合物水热转化过程中可溶性副产物的鉴定 157

  5.5.1 HMF 水热转化生成的可溶性副产物鉴定 157

  5.5.2 糠醛水热转化生成的可溶性物质鉴定 160

  5.5.3 糠醇水热转化生成的可溶性物质鉴定 162

  5.5.4 呋喃类物质的水热转化路径 166

 5.6 C—C 键水解断裂反应 169

 5.7 α-羰基醛的生成与转化路径 170

 5.8 碳水化合物的降解路径 171

 5.9 结论 174

 参考文献 175

附录

180 葡萄糖及生物质衍生呋喃类物质水热转化生成的可溶性物质的 MS 和 MS^2 数据及谱图

第一章

概　　述

1.1 生物质和木质纤维素的定义

在某些文献中,将生物质定义为一切来源于植物、动物和微生物等生命体的有机物质。根据这一定义,含有蔗糖、淀粉、动植物油脂等可供人类食用的物质也包含于生物质的概念之中。为了避免将蔗糖、淀粉、动植物油脂等可食用物质用于生产能源和化学品所带来的"与人争粮"的问题,通常将生物质定义为:来源于生命体且在人类生产和生活过程中当作废弃物的有机物质。在这种定义之下,生物质主要包括农作物秸秆、林业废弃物、工业有机废弃物(造纸黑液和酒糟)、藻类、厨房垃圾、污水处理厂剩余污泥和人畜粪便等。

上述多种生物质资源中,以农作物秸秆和林业废弃物为代表的木质纤维素具有相对较单一的结构,其主要成分为纤维素、半纤维素和木质素,在利用过程中所涉及的物理变化和化学反应具有明显的规律。与其他种类的生物质相比,木质纤维素储量巨大、存在广泛。据统计,自然界每年大约生产 1.8×10^3 亿吨木质纤维素[1,2]。因此,深入研究木质纤维素类生物质资源的利用技术,对人类社会的可持续发展具有重要意义。

本书讨论对象集中于以农业废弃物和林业废弃物为代表的木质纤维素类生物质,如麦秸秆、稻草、稻壳、玉米秸秆、甘蔗渣、专用能源作物、木材加工废料、薪柴和森林废弃物等。其他种类的生物质资源,如动植物油脂、工业有机废水、生活污水、人畜粪便等不在本书讨论范围之内。

1.2 木质纤维素的种类及其化学结构

1.2.1 木质纤维素类的种类

木质纤维素主要来源于植物体的茎干、叶子和果实。自然界中存在的植物种类繁多,根据植物的形态特征及其化学成分差异,通常将植物分为三类:针叶材植物、阔叶材植物和禾本科植物[3]。以这三类植物的茎干为原料所获得的木质纤维素,分别称之为针叶材木质纤维素、阔叶材木质纤维素和禾本科木质纤维素。这三类木质纤维素在结构上存在一定的差异。

(1)针叶材木质纤维素

由于这类木质纤维素植物原料的叶子多呈针叶形、条形或鳞形,故一般称之为针叶材木质纤维素。同时，因为这类木质纤维素的材质一般比较松软，故其植物原料又被称为软木（soft wood）。常见的针叶材有云杉、冷杉、马尾松、落叶松、湿地松、火炬松等。

（2）阔叶材木质纤维素

这类木质纤维素植物原料的叶子多为宽阔状，故称阔叶材木质纤维素。由于这类木质纤维素的材质较为坚硬，故其植物原料又被称为硬木（hard wood）。常用的阔叶材有杨木、桦木、桉木、榉木、楹木、相思木等。

（3）禾本科木质纤维素

禾本科木质纤维素来源于禾本科植物（草本植物）的茎干。在我国分布广泛，种类繁多，比如甘蔗渣、高粱秆、玉米秸秆、稻草、麦秸秆、芦苇、荻、芦竹、芒秆等。我国每年的农业生产过程中都会产生大量的农作物秸秆，如玉米秸秆、稻草、麦秸秆等，这些秸秆资源的收集困难且缺乏有效的利用技术，所以大量的秸秆资源常被农户就地焚烧，既导致资源的浪费，又导致环境的污染。对秸秆资源进行高值化利用，既能提高农作物产品的经济价值，又能够解决就地焚烧秸秆所带来的环境污染问题，对提高农村人口的经济水平和改善农村地区的环境具有重要意义。

1.2.2 木质纤维素的化学结构

木质纤维素是由多种复杂的高分子有机化合物组成的复合体，其化学组成主要有纤维素、半纤维素、木质素、提取物和灰分等。其中，纤维素、半纤维素和木质素是木质纤维素类生物质的主要化学成分，这三种组分是构成植物细胞壁的主要成分。如图 1-1 所示，纤维素组成微纤丝，微纤丝则是构成植物细胞壁的网状骨架，而半纤维素和木质素则是填充在微纤丝之间的"黏合剂"和"填充剂"。在一般的植物纤维原料中，这三种成分的质量占原料总质量的 80%~95%。从更微观的分子水平角度来讲，纤维素是由葡萄糖单元通过 β-1,4-糖苷键链接而成的多糖，半纤维素是由五碳糖（木糖、阿拉伯糖等）和六碳糖（葡萄糖、甘露糖、半乳糖、鼠李糖等）组成的多糖，而木质素是由香豆醇、针叶醇和芥子醇这三种苯基丙烷结构单元组成的高分子聚合物。

通常情况下，纤维素的含量约占植物总质量的 35%~50%，其次是半纤维素约占 25%~35%，剩下是木质素约占 15%~30%[6,7]。表 1-1 列出了常见植物中纤维素、半纤维素和木质素含量。可以看到，不同的原料中纤维素、半纤维素和木质素的含量有所不同。一般植物茎干中的纤维素含量最高，而半纤

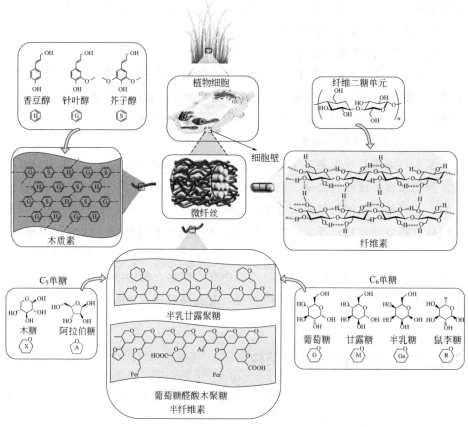

图 1-1 木质纤维素的组成及各类组分的代表结构[4,5]

维素含量次之,也有少量植物的木质素含量高达30%以上。禾本科植物中的木质素含量明显低于软木和硬木中的木质素含量。需要说明的是,各类植物的生长环境、生长年龄等对其化学组成有一定的影响,且同一株植物在不同部位的化学组成也有区别,所以,表1-1所展示的含量仅能作为参考。

表 1-1 常见植物中纤维素、半纤维素和木质素含量[8-12]

种类	名称	纤维素/%	半纤维素/%	木质素/%
软木	冷杉	38.8	29.1	28.5
	黄杉	38.8	29.3	26.3
	铁杉	37.7	30.5	27.9
	刺柏	33.0	32.1	30.3
	辐射松	37.4	27.2	33.2

续表

种类	名称	纤维素/%	半纤维素/%	木质素/%
软木	苏格兰松	40.0	27.7	28.5
	挪威云杉	41.7	27.4	28.3
	白云杉	39.5	27.5	30.6
	落叶松	41.4	26.8	29.6
硬木	红枫	42.0	28.9	25.4
	糖枫	40.7	30.8	25.2
	颤杨	49.4	28.3	18.1
	山毛榉	39.4	33.3	24.8
	银桦	41.0	32.4	22.0
	纸桦	39.4	34.5	21.4
	赤杨	38.3	19.2	24.8
	赤桉	45.0	24.4	31.3
	蓝桉	51.3	25.2	21.9
	黑荆树	42.9	33.6	20.8
禾本科植物	玉米秸秆	37.5	22.4	17.6
	玉米芯	33.7	31.9	6.1
	柳枝稷	39.5	25.0	17.8
	麦壳	33.6	37.2	19.3
	大麦秸秆	33.8	21.9	13.8
	小麦秸秆	30.2	22.3	17.0
	稻草	31.1	22.3	13.3
	甘蔗渣	43.1	31.1	11.4
	甜高粱秸秆	27.3	14.5	14.3

1.2.2.1 纤维素

纤维素（cellulose）是植物纤维中最主要的化学成分，也是自然界中最为丰富的高分子碳水化合物。如图 1-2 所示，纤维素由 D-吡喃葡萄糖单元通过 β-1,4-糖苷键线型连接而成，其分子式为 $(C_6H_{10}O_5)_n$，其中 n 为聚合度。纤维素单体间的线性排列及高度有序性导致在纤维素链之间存在大量的氢键，进而导致纤维素结晶度高、比表面积小、结构稳定，使得其难以在温和条件下水解。

纤维素同时含有结晶区和无定形区[2]。在水解过程中，无定形区的纤维素能被迅速水解，而结晶区纤维素则由于 β-1,4-糖苷键难与水及催化剂（或水解酶）接触而难以被水解。为了将结晶纤维素转变为无定形纤维素从而提高反

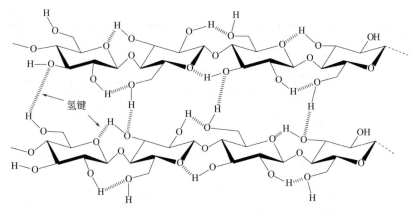

图 1-2　纤维素的分子结构示意图

应活性，人们常采用物理处理（球磨、研磨、蒸煮、气爆等）和化学处理（碱处理、酸处理、氧化处理等）来降低聚合度，增加还原端基团的数量。特别地，用磷酸处理纤维素可以增大其可接触的表面积，虽然在此过程中结晶纤维素的聚合度并没有下降。

纤维素分子中葡萄糖单元的含量即为纤维素分子的聚合度。不同木质纤维素原料之间以及同一植物的不同部位，纤维素的聚合度大小有所不同。表 1-2 列出了一些代表性木质纤维素原料中纤维素的聚合度。初生细胞壁中纤维素平均聚合度约为 6000 个葡萄糖单元，而在次生细胞壁中可达 14000 个。相对而言，棉花、麻类的纤维素聚合度较高，木材类生物质的聚合度次之，而麦秸秆、甘蔗渣等的纤维素平均聚合度则较低。

表 1-2　一些代表性木质纤维素原料中的纤维素的聚合度[13-15]

原料种类	聚合度[葡萄糖单元]
生棉花	7000
纯化棉	300～1500
短绒棉	3170
亚麻	8000
苎麻	6500
云杉纸浆	3300
山毛榉	4050
红枫	4450
白云杉	4000
银桦	5500
颤杨	5000

续表

原料种类	聚合度[葡萄糖单元]
细菌纤维素	2700
辐射松	3063
滤纸	1500
玉米秸秆	1000
麦秸秆	1045
甘蔗渣	925
醋酸纤维素	600

在木质纤维素水热炼制工艺中，纤维素解聚为葡萄糖，而葡萄糖则进一步被选择性地转化为多种平台化学品，如5-羟甲基糠醛、乙酰丙酸、乳酸、山梨醇、乙二醇、1,2-丙二醇等。纤维素的水热催化转化是本书的重点，在第二章将详细论述纤维素催化转化过程的基本原理。

1.2.2.2 半纤维素

半纤维素（hemicellulose）是在植物细胞壁中除了纤维素、果胶和淀粉之外的全部碳水化合物的统称，也被称为非纤维素碳水化合物，是木质纤维素中第二种大分子碳水化合物。在植物细胞壁中，半纤维素是填充在微纤丝之间的"黏合剂"和"填充剂"。

与纤维素不同，半纤维素是一类聚合度在70～200之间的高度分支的杂多糖，由一系列亚单位组成，这些亚单位包括糖、糖酸和非碳水化合物基团。这种复杂的结构将半纤维素和植物细胞壁中其他多糖区分开，赋予它们水溶性、易吸水性，并与植物细胞壁许多其他成分高度交联。构成半纤维素线性聚糖主链的单糖主要是葡萄糖、木糖、甘露糖和半乳糖；而构成半纤维素侧链的糖基有木糖、葡萄糖、半乳糖、阿拉伯糖、岩藻糖、鼠李糖和葡萄糖醛酸、半乳糖醛酸等。木糖、甘露糖、半乳糖、阿拉伯糖和鼠李糖的分子结构见图1-3。其中，木糖和阿拉伯糖是五碳醛糖，而半乳糖、甘露糖和鼠李糖是六碳醛糖。

针叶材、阔叶材和禾本科植物中的半纤维素的化学成分有明显的区别，表1-3列出了各类植物中半纤维素的主要成分及含量[2,11,16,17]。在针叶木植物的茎干中，半纤维素主要由半乳葡甘露聚糖（约为半纤维素总质量的60%～70%）和阿拉伯甲基葡糖醛酸木聚糖（约为半纤维素总质量的15%～30%）组成。在阔叶材植物中，最重要的半纤维素成分是木聚糖，尤其是葡萄糖醛酸木聚糖，约占阔叶材中半纤维素的80%～90%。在禾本科植物中，半纤维素主要成分为木聚糖。可以看到，葡萄糖醛酸木聚糖是阔叶材植物和禾本

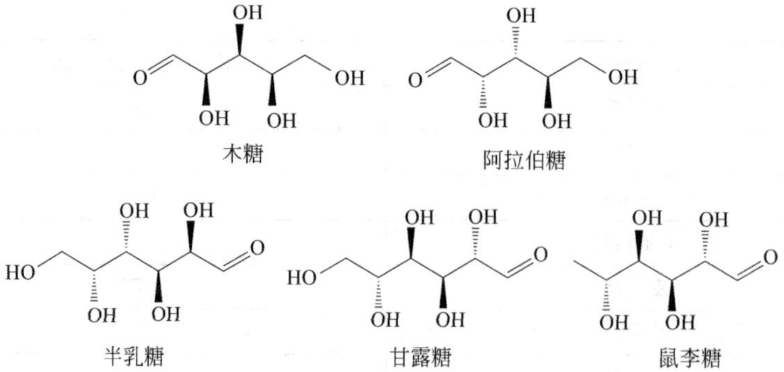

图 1-3 几种半纤维素中单糖的分子结构

科植物中半纤维素的重要组成部分。图 1-4 展示了葡萄糖醛酸木聚糖的分子结构。

表 1-3 各类植物中半纤维素的主要成分及含量[2,11,12]

半纤维素组分	针叶材植物/%	阔叶材植物/%	禾本科植物/%
葡萄糖醛酸木聚糖	20～45	80～90	>90
葡甘露聚糖	1～5	1～5	0
半乳葡甘露聚糖	60～70	0.1～1	0
阿拉伯半乳聚糖	1～15	0.1～1	1～5
其他半乳聚糖	0.1～1	0.1～1	1～5

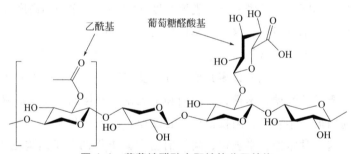

图 1-4 葡萄糖醛酸木聚糖的分子结构

在木质纤维素水热炼制过程中，半纤维素可以在相对温和的条件下水解生成单糖，从而实现半纤维素与其他组分的分离。特别地，禾本科植物和阔叶材植物的半纤维素富含木聚糖，因此这些生物质的半纤维素在生物质水热炼制中可以水解为木糖，并进而转化为糠醛[17]。此外，戊糖可用于生产结晶木糖，进而催化加氢生产木糖醇，或用硝酸氧化生产糖酸（如木糖酸和三羟基戊二酸等）。需要说明的是，因为半纤维素的组成和结构是比较多样化的，所以其反应情况比纤维素更加复杂，反应产物的化学性质也有着一定的差异。

1.2.2.3 木质素

木质素（lignin）作为细胞间固结物质填充在细胞壁的微纤丝之间，也存在于胞间层中，是含量仅次于纤维素的一种大分子有机物质。在植物细胞中，木质素像天然胶水一样将纤维素和半纤维素紧密地黏结在一起，有着加固木质化植物组织的作用，木质化后的细胞壁可以增大树木茎干的硬度，大大提高木质纤维素的力学性能和抵抗微生物和病原体的能力。

（1）木质素的结构单元

木质素是具有三维空间结构的高分子，其基本结构单位为苯基丙烷单元。在苯基丙烷的苯环上可能有0个、1个或2个甲氧基，分别称为对羟基苯基丙烷结构（H型）、愈创木基丙烷结构（G型）和紫丁香基丙烷结构（S型）[8]。由于木质素中的苯基丙烷单元又是由香豆醇、针叶醇和芥子醇这三种初始前驱体聚合而成，所以部分文献也把香豆醇、针叶醇和芥子醇称为木质素的单体。如表1-4所示，木质素及其初级单体之间的比例在不同的植物之间存在明显的不同。木质素在各类生物质中的含量按以下顺序递减：针叶材＞阔叶材＞禾本科植物。针叶材和禾本科植物中的木质素主要由松柏醇单体（愈创木基结构）构成，而阔叶材的木质素单体包含了50%的松柏醇单体和50%的芥子醇单体（丁香基结构）[18,19]。此外，禾本科植物的木质素中含有7%～12%的由香豆酸和阿魏酸形成的酯基。相比于针叶材和阔叶材，禾本科植物的木质素分子量低，分散性好，在分离过程中易溶出。

表1-4 木质素中单体结构及其在各类生物质中的分布[19]

木质纤维素种类	木质素含量/%	单体种类及含量/%		
		香豆醇	松柏醇	芥子醇
针叶材	27～33	0	90～95	5～10
阔叶材	18～25	0	50	50
禾本科植物	17～24	5	75	25

因为苯基丙烷单元是木质素的基本单元，且每个单元中都含有一定数量的甲氧基，所以在表示木质素的元素分析结果时，常用除去甲氧基量的苯丙

(C_6—C_3) 单元做标准，以相当于 C_9 的各种元素量来表示，再加上相当于每个 C_9 的甲氧基数。几种常见植物的磨木木质素的 C_9 单元见表 1-5。

表 1-5　几种常见植物的磨木木质素的 C_9 单元

磨木木质素来源	C_9 单元式	磨木木质素来源	C_9 单元式
云杉	$C_9H_{8.83}O_{2.37}(OCH_3)_{0.96}$	稻草	$C_9H_{7.44}O_{3.38}(OCH_3)_{1.03}$
山毛榉	$C_9H_{7.10}O_{2.41}(OCH_3)_{1.36}$	芦竹	$C_9H_{7.81}O_{3.12}(OCH_3)_{1.18}$
桦木	$C_9H_{9.03}O_{2.77}(OCH_3)_{1.58}$	甘蔗渣	$C_9H_{7.34}O_{3.50}(OCH_3)_{1.10}$
麦秸秆	$C_9H_{7.39}O_{3.00}(OCH_3)_{1.07}$	毛竹	$C_9H_{7.33}O_{3.81}(OCH_3)_{1.24}$

（2）木质素结构单元的连接方式

在天然木质素中，2/3 或更多的木质素单体通过醚键连接，而其他单体的连接键是 C—C 键。醚键主要是苯基丙烷单元中苯环上的酚羟基与侧链上的羟基形成的酚醚键及侧链上的羟基之间形成的烷醚键。为了对两个单体之间各种键的类型进行分类，通常将木质素单体的脂肪族侧链中的碳原子标记为 α、β 和 γ，而芳香族部分中的碳原子则标记为 1~6（如表 1-4 中的香豆醇结构）。例如，β—O—4 键表示脂肪族侧链的 β 碳和芳香族部分的 C_4 位置上的氧原子之间形成的键（图 1-5，第一结构模型）。图 1-5 展示了木质素结构单元之间的主要键合方式。可以看到，木质素单体之间主要通过 β—O—4（β-芳基醚）、β—β（树脂）和 β—5（苯基香豆满）连接。其他键合方式包括 α—O—4（α-芳基醚）、4—O—5（二芳醚）、5—5、α—O—γ（脂肪醚）等[20,21]。表 1-6 则列出了木质素中的主要连接键和官能团的数量。可以看到，不论是针叶材还是阔叶材，β—O—4（β-芳基醚）都是木质素单体连接的重要方式。应该注意的是，即使对于相同的植物种类，由于生长环境、生长年龄甚至分析方法等因素，这些数据也有很大的差异。

表 1-6　木质素中的主要连接键和官能团的数量[20]

连接键	每 100 个苯丙单元中的数量		官能团	每 100 个苯丙单元中的数量	
	针叶材	阔叶材		针叶材	阔叶材
β—O—4	43~50	50~65	甲氧基	92~96	132~146
β—5	9~12	4~6	酚羟基	20~28	9~20
α—O—4	6~8	4~8	苯族羟基	16	—
β—β	2~4	3~7	脂肪族羟基	120	—
5—5	10~25	4~10	羰基	20	3~17
4—O—5	4	6~7	羧基	—	11~13
β—1	3~7	5~7			
其他	16	7~8			

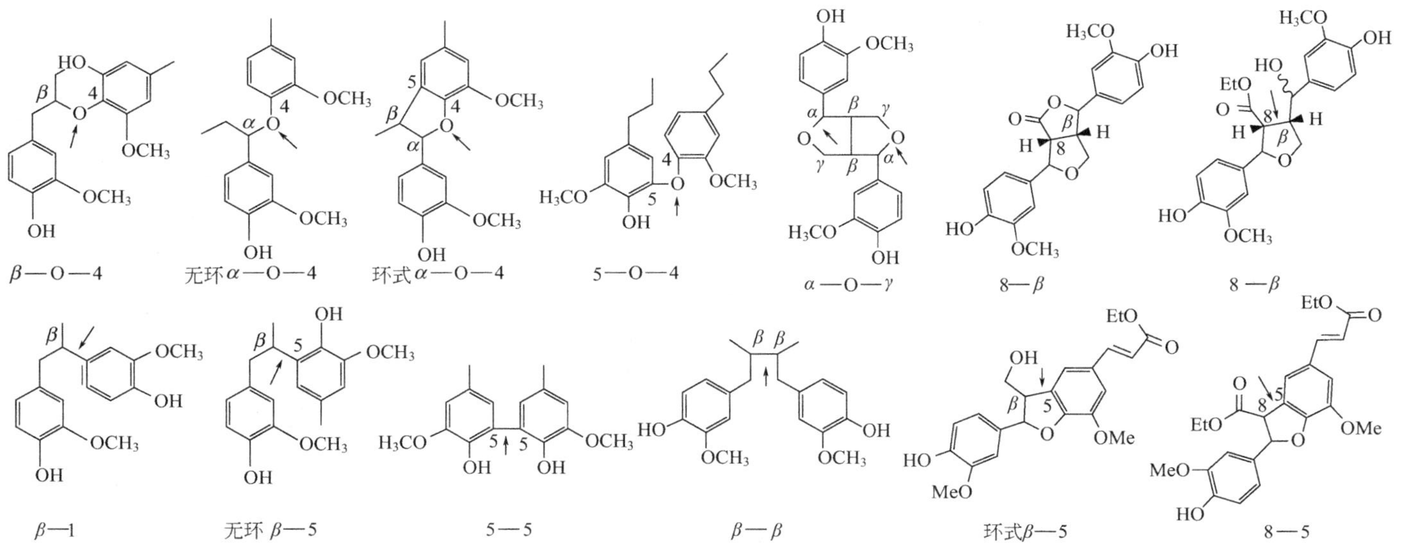

图 1-5 木质素结构单元之间的主要键合方式[20, 22]

图1-6展示了典型的木质素分子结构模型。这个模型不是通常意义上的木质素的精确结构式，而是一个用来说明木质素单体及单体间连接方式的模型。单体之间的键基本上决定了木质素的反应活性。由于 β—O—4（β-芳基醚）是木质素中最常见的键，其化学反应活性在很大程度上决定了木质素对化学降解的抵抗力。影响反应活性的另一个重要因素是官能团，包括甲氧基、苯甲醇、酚羟基、非环苄基醚、羧基和羰基等[20]。

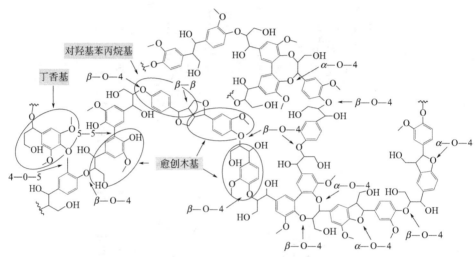

图1-6 典型的木质素分子结构模型[20]

由于木质素分子结构中含有较多的烷基酚单体，且其结构中存在芳香基、酚羟基、醇羟基、羰基、甲氧基、羧基、共轭双键等活性基团，可以进行氧化、还原、水解、醇解、光解、酰化、磺化、烷基化、卤化、硝化、缩合和接枝共聚等化学反应。在木质纤维素炼制中，木质素被认为是生产可挥发酚类、芳烃和新型碳材料的原料[22,23]。

1.3 木质纤维素的利用

以木质纤维素为资源能够获取的产品包括热能、燃料和化工原料。目前国内外开发了多种木质纤维素的利用技术，如图1-7所示。以获取热能为目的的木质纤维素利用技术主要为直接燃烧技术；以获取固体燃料为目的的木质纤维素利用技术包括压缩成型技术、热解技术、水热碳化技术等；以获取气体燃料或气体化工原料为目的的木质纤维素利用技术包括气化技术、热解技术、沼气发酵技术等；以获取液体燃料或液体化工原料的木质纤维素利用技术包括气化-间接液化技术、热解技术、水热液化技术、水解-发酵技术等[24,25]。

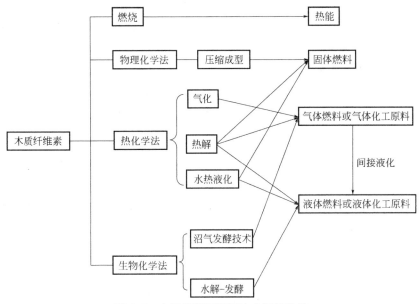

图 1-7　木质纤维素利用技术及其产品

在上述多种木质纤维素转化所获得的产品中，液体燃料及液体化工原料由于与石油衍生产品性质相似，其后续加工炼制较为容易，因此具有相对较高的附加值，是国内外的研究热点[26,27]。如图 1-7 所示，利用木质纤维素制备液体燃料或化工原料的技术，根据过程中涉及的温度、压力和反应介质不同，可以分为木质纤维素气化-间接液化技术、木质纤维素热解技术、木质纤维素水热炼制技术和木质纤维素水解-发酵技术。

（1）木质纤维素气化-间接液化技术

该技术与当前工业化的煤间接液化技术相似，首先在高温条件下、采用水蒸气和 O_2 对木质纤维素进行气化，得到以 CO、H_2 和 CH_4 为主要成分的气体产物，然后以这些气体产物为原料，通过费-托合成、甲醇合成、烯烃合成等技术，得到以烷烃、烯烃、甲醇等为主要成分的液体产物。

（2）木质纤维素热解技术

该过程的转化温度一般在 400℃ 以上，木质纤维素中的主要成分直接转化成气体混合物（10%～40%）、黏稠的油状液体（30%～75%）和固体残渣（10%～58%）。由于热解油的成分极为复杂且含有大量的呋喃环、酚环、羟基、羧基、羰基等官能团，热解油的稳定性较差，且热值较低，难以直接用作液体燃料，仍然需要通过加氢脱氧处理，才能替代液体燃料使用。热解油的加氢脱氧处理过程一般以 Pt、Pd、Ni、Fe、Cu、Co 等金属负载于固体酸载体上得到的加氢催化剂，在 300~600℃ 的高温和 2~20MPa 的 H_2 压力条件下，将生物油中的氧元素进行脱除，最终获得氧元素含量小于 0.7%（质量分数）的稳定性较好的生物油。

（3）木质纤维素水热炼制技术

该技术也是一种热化学转化过程，但是转化温度要比热裂解温度低很多。该过程采用液体（水、甲醇、乙醇、苯酚等）为反应介质，一般在 130~300℃ 的温度下，将木质纤维素中的纤维素、半纤维素和木质素组分通过一系列分解和脱水反应，选择性地转化为小分子含氧有机物，包括 5-羟甲基糠醛、糠醛、乙酰丙酸（酯）、乳酸（酯）、山梨醇、木糖醇、1,2-丙二醇、乙二醇、甲酸、苯酚等，然后对这些小分子含氧有机物进行酯化、缩合、加氢脱氧等反应，最终得到性质稳定、易挥发的液体有机物[28,29]。为了实现木质纤维素组分的定向转化，在木质纤维素水热炼制过程中通常会加入各种各样的催化剂，而催化剂的制备成本、催化剂的活性和回收过程是该类技术最终走向应用的关键。此外，木质纤维素水热炼制过程涉及木质纤维素解聚、平台化学品分离和加工等步骤，且各步骤之间的催化剂不同，会导致该类技术的分离成本较高。

（4）木质纤维素水解-发酵技术

该技术在微生物的作用下将木质纤维素中的纤维素、半纤维素和木质素等组分转化为液体有机物（乙醇、丁醇、琥珀酸等）。利用淀粉、蔗糖等易水解的原料发酵制备乙醇和丁醇的技术已经很成熟，但是以木质纤维素为原料通过发酵制备乙醇和丁醇的技术距离工业化仍有较大距离，这是因为木质纤维素中的纤维素组分难以通过低成本的手段水解得到能够用于发酵的水解液[12,30-32]。

鉴于本书的出版目的，本书在后面的章节中将重点叙述木质纤维素的水热炼制原理及技术，其他木质纤维素转化技术将不再是我们的讨论对象。

参 考 文 献

[1] Li X, Xu R, Yang J, Nie S, Liu D, Liu Y, Si C. Production of 5-hydroxymethylfurfural and levulinic acid from lignocellulosic biomass and catalytic upgradation. Industrial Crops and Products, 2019, 130: 184-197.

[2] MäKi-Arvela P I, Salmi T, Holmbom B, Willfö R S, Murzin D Y. Synthesis of sugars by hydrolysis of hemicelluloses—a review. Chemical Reviews, 2011, 111 (9): 5638-5666.

[3] 杨淑蕙. 植物纤维化学. 北京：中国轻工业出版社, 2011.

[4] Isikgor F H, Becer C R. Lignocellulosic biomass: a sustainable platform for the production of bio-

based chemicals and polymers. Polymer Chemistry, 2015, 6 (25): 4497-4559.

[5] 聂根阔. 基于松节油和木质纤维素平台化合物的高密度燃料合成. 天津: 天津大学, 2017.

[6] Alonso D M, Bond J Q, Dumesic J A. Catalytic conversion of biomass to biofuels. Green Chemistry, 2010, 12 (9): 1493-1513.

[7] Mckendry P. Energy production from biomass (part 1): overview of biomass. Bioresource Technology, 2002, 83: 37-46.

[8] Malherbe S, Cloete T E. Lignocellulose biodegradation: fundamentals and applications. Reviews in Environmental Science and Bio/Technology, 2002, 1: 105-114.

[9] Mosier N, Wyman C, Dale B, Elander R, Lee Y Y, Holtzapple M, Ladisch M. Features of promising technologies for pretreatment of lignocellulosic biomass. Bioresource Technology, 2005, 96 (6): 673-686.

[10] Sun Y, Cheng J. Hydrolysis of lignocellulosic materials for ethanol production: a review. Bioresource Technology, 2002, 83: 1-11.

[11] Amidon T E, Wood C D, Shupe A M, Wang Y, Graves M, Liu S. Biorefinery: conversion of woody biomass to chemicals, energy and materials. Journal of Biobased Materials and Bioenergy, 2008, 2 (2): 100-120.

[12] Haghighi Mood S, Hossein Golfeshan A, Tabatabaei M, Salehi Jouzani G, Najafi G H, Gholami M, Ardjmand M. Lignocellulosic biomass to bioethanol, a comprehensive review with a focus on pretreatment. Renewable and Sustainable Energy Reviews, 2013, 27: 77-93.

[13] Rinaldi R, Schuth F. Acid hydrolysis of cellulose as the entry point into biorefinery schemes. ChemSusChem, 2009, 2 (12): 1096-1107.

[14] Puri V P. Effect of crystallinity and degree of polymerization of cellulose on enzymatic saccharification. Biotechnology and Bioengineering, 1984, 26 (10): 1219-1222.

[15] Kumar R, Mago G, Balan V, Wyman C E. Physical and chemical characterizations of corn stover and poplar solids resulting from leading pretreatment technologies. Bioresource Technology, 2009, 100 (17): 3948-3962.

[16] Amidon T E, Liu S. Water-based woody biorefinery. Biotechnology Advances, 2009, 27 (5): 542-550.

[17] Liu S J, Lu H F, Hu R F, Shupe A, Lin L, Liang B. A sustainable woody biomass biorefinery. Biotechnology Advances, 2012, 30 (4): 785-810.

[18] Zakzeski J, Bruijnincx P C A, Jongerius A L, Weckhuysen B M. The catalytic valorization of lignin for the production of renewable chemicals. Chemcal Reviews, 2010, 110 (6): 3552-3599.

[19] Azadi P, Inderwildi O R, Farnood R, King D A. Liquid fuels, hydrogen and chemicals from lignin: a critical review. Renewable and Sustainable Energy Reviews, 2013, 21: 506-523.

[20] Li C, Zhao X, Wang A, Huber G W, Zhang T. Catalytic Transformation of lignin for the production of chemicals and fuels. Chemical Reviews, 2015, 115 (21): 11559-11624.

[21] 舒日洋, 徐莹, 张琦, 马隆龙, 王铁军. 木质素催化解聚的研究进展. 化工学报, 2016, 67 (11): 4523-4532.

[22] Pandey M P, Kim C S. Lignin Depolymerization and conversion: a review of thermochemical

methods. Chemical Engineering and Technology, 2011, 34 (1): 29-41.
[23] Zhu Y, Li Z, Chen J. Applications of lignin-derived catalysts for green synthesis. Green Energy and Environment, 2019, 4 (3): 210-244.
[24] Nanda S, Mohammad J, Reddy S N, Kozinski J A, Dalai A K. Pathways of lignocellulosic biomass conversion to renewable fuels. Biomass Conversion and Biorefinery, 2013, 4 (2): 157-191.
[25] Gallezot P. Process options for converting renewable feedstocks to bioproducts. Green Chemistry, 2007, 9 (4): 295-302.
[26] Huber G W, Iborra S, Corma A. Synthesis of transportation fuels from biomass: Chemistry, catalysts, and engineering. Chemical Reviews, 2006, 106 (9): 4044-4098.
[27] Zhang Z, Song J, Han B. Catalytic transformation of lignocellulose into chemicals and fuel products in ionic liquids. Chemical Reviews, 2017, 117 (10): 6834-6880.
[28] Besson M, Gallezot P, Pinel C. Conversion of biomass into chemicals over metal catalysts. Chemical Reviews, 2014, 114 (3): 1827-1870.
[29] Ragauskas A J, Beckham G T, Biddy M J, Chandra R, Chen F, Davis M F, Davison B H, Dixon R A, Gilna P, Keller M, Langan P, Naskar A K, Saddler J N, Tschaplinski T J, Tuskan G A, Wyman C E. Lignin valorization: improving lignin processing in the biorefinery. Science, 2014, 344 (6185): 1246843-1246843.
[30] Sarkar N, Ghosh S K, Bannerjee S, Aikat K. Bioethanol production from agricultural wastes: an overview. Renewable Energy, 2012, 37 (1): 19-27.
[31] Binod P, Sindhu R, Singhania R R, Vikram S, Devi L, Nagalakshmi S, Kurien N, Sukumaran R K, Pandey A. Bioethanol production from rice straw: an overview. Bioresource Technology, 2010, 101 (13): 4767-4774.
[32] Talebnia F, Karakashev D, Angelidaki I. Production of bioethanol from wheat straw: an overview on pretreatment, hydrolysis and fermentation. Bioresource Technology, 2010, 101 (13): 4744-4753.

第二章

纤维素水热催化转化制备平台化学品的基本原理

2.1 引言

碳水化合物（包括纤维素和半纤维素）是木质纤维素的主要成分，故人们非常关注这些碳水化合物组分的转化。相比于半纤维素，纤维素在自然界中的储量更高，且其降解难度更大，因此，纤维素的水热催化转化在木质纤维素水热转化制备生物质基液体燃料及化学品中占据重要地位。

如图2-1所示，在水热条件下，纤维素可发生水解生成葡萄糖，而葡萄糖则可以经一系列反应转化为多种小分子含氧化合物，如5-羟甲基糠醛（HMF）、乙酰丙酸（酯）、甲酸、乳酸（酯）、山梨醇、乙二醇、1,2-丙二醇等[1-5]。这些小分子含氧化合物可以作为化工原料，用于生产生物质基液体燃

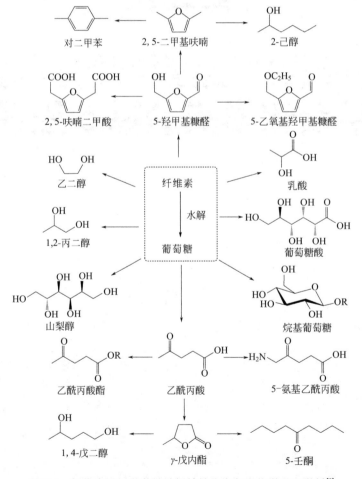

图2-1 纤维素通过葡萄糖选择性转化为各类化学品和燃料[6]

料、有机材料、有机溶剂等，从而降低对不可再生的化石燃料的依赖。

国内外在纤维素水热催化转化领域开展了广泛而深入的研究，开发出了多种反应体系以实现对纤维素高效、高选择性地转化。本章主要对葡萄糖和纤维素转化为各类含氧平台化合物的反应路径及反应原理进行介绍。

2.2 纤维素的催化水解反应

纤维素的水解反应被认为是生物精炼的起点，在纤维素选择性转化为目标产物的整个过程中占据重要地位。国内外对纤维素的水解技术与基础理论开展了大量的研究工作[6-8]，本节主要从反应介质和催化剂两个方面对部分具有代表性的技术进行介绍。

2.2.1 纤维素催化水解的反应介质

水是纤维素水解最理想也最常用的溶剂。在无催化剂的情况下，纤维素可于220℃以上的热水中发生水解[9-11]。特别地，在300℃以上的高温蒸汽中，纤维素可以迅速地转化为无定形状态并发生水解[12]。因此，在不添加催化剂的情况下，木质纤维素水热转化过程通常都需要在220℃以上的温度下才能顺利进行。比如，木质纤维素的水热碳化过程通常都需要在220℃以上的温度下进行[13-15]，而木质纤维素水热液化制备生物油的过程通常都需要在250℃的温度下进行[16,17]。当在220℃以上的水热环境中处理纤维素时，即使催化剂的酸性较弱，纤维素也能够较为顺利地转化。比如，在245℃的水热条件下催化纤维素水热转化生成乙二醇时，即使催化剂（如WO_3）的酸性很弱，纤维素也能够在较短的时间内被转化[18]。

离子液体是一类在常温下呈液态的溶剂。由于离子液体能够溶解纤维素，从而促进纤维素的水解反应，所以采用离子液体作纤维素水解的介质时，纤维素的水解反应可以在较为温和的条件下进行[3]。近年来，离子液体在木质纤维素转化为化学品和燃料产品中的应用引起了广泛的研究兴趣。然而，离子液体的合成成本较高且对环境具有潜在的危害，所以采用离子液体作反应介质时会涉及溶剂回收、产物分离等，同样会增加整个工艺的生产成本。

2.2.2 纤维素催化水解的催化剂

添加催化剂可以让纤维素的水解反应在更温和的条件下进行并得到更多的

目标产物，所以，纤维素水解催化剂的研发是目前纤维素水解的研究热点。纤维素水解可以分为酶催化和酸催化两大类。酶催化水解在非常温和的条件下进行，但是纤维素水解酶价格相对昂贵，限制了酶水解技术的大规模工业化应用。本书主要涉及化学催化手段转化纤维素。常用于纤维素水解的酸催化剂包括无机酸（如 H_2SO_4、H_3PO_4、HCl 等）、能溶于水的酸性金属盐 [如 $NaHSO_4$、$CuCl_2$、$AlCl_3$、$CrCl_3$、$SnCl_2$、$ZnCl_2$、$Al_2(SO_4)_3$、$Fe_2(SO_4)_3$][7,8,19,20] 或者固体酸（如磺化碳材料、磺化树脂、酸性复合金属氧化物、酸性分子筛等）[8]。

（1）均相无机酸催化剂

无机酸（HCl、H_2SO_4 和 H_3PO_4 等）是纤维素水解反应中最常用的催化剂。这些无机酸由于价格便宜、易于获取，目前仍然是最具应用前景的纤维素水解催化剂。

无机酸催化水解纤维素具有较为悠久的历史和成熟的水解工艺，可追溯至 19 世纪初期，且在 20 世纪初就达到了工业化生产水平。无机酸催化纤维素水解的工艺又可分为浓酸水解和稀酸水解两种。

浓酸以其独特的性质进入到纤维素分子内部，打破氢键的同时使纤维素产生润胀效应，随后纤维素大分子链中的糖苷键在浓酸中的质子催化作用下断裂，生成葡萄糖、纤维二糖和其他低聚糖等中间体。较低的温度下，纤维素在 41%～42%的浓 HCl、65%～70%的浓 H_2SO_4 或 80%～85%的浓 H_3PO_4 中发生均相水解反应均可得到含有葡萄糖和低聚糖的水解液，再经稀释加热即可进一步将水解液中的低聚糖水解为葡萄糖。浓酸催化木质纤维素水解能够得到较高的单糖收率。实际上，目前实验室分析木质纤维素结构中多糖的标准方法，即根据浓酸水解纤维素能够得到较高单糖收率的原理。室温下，将样品用 72%（质量分数） H_2SO_4 处理 1h，将多糖全部转化为可溶性糖，随后用水稀释至 4% H_2SO_4，加热至 121℃并保持 1h，从而将低聚糖进一步水解为单糖。有研究发现，使用 7∶3 的 H_3PO_4 和 H_2SO_4 混合物可以显著提高半纤维素糖在酸中的稳定性，能通过简单地将催化剂和生物质原料加热到 85℃保持 4h，以高固相反应 [20%（质量分数）固相] 获得可溶性碳水化合物的高收率（80%～90%）。浓酸水解具有反应条件温和（对温度和压力要求低）、能耗低、葡萄糖收率高的优点，但其弊端是对原材料的含水量和设备的防腐蚀性能要求严格，同时浓酸的使用会严重污染环境。目前开发出的浓酸水解工艺包括 Bergius 工艺、Hoechst 工艺等，采用的催化剂包括浓 H_3PO_4、浓 H_2SO_4、浓 HCl、无水 HF 等。需要指出的是，酸催化剂的回收是浓酸催化木质纤维素水解工艺是能否成功工业化应用的关键因素,而 H_2SO_4 和 H_3PO_4 的沸点分别为 337℃ 和 158℃,使得它们无法通过蒸发回收。由于 HCl 和 HF 的挥发性较强,可以通过蒸发进行回收,所以采用这两种催化剂的浓酸水解工艺相对较具经济性。但是,HCl 和 HF 回收过程中的能耗高及回收催化剂的量较少同样使得这些过程不经济。

稀酸中大量的水合氢离子会进攻纤维素的糖苷键使得其氧原子发生质子化,此时糖苷键较为活泼并且易发生断裂,进而纤维素被水解为小分子糖类。与浓酸水解纤维素相比,稀酸能有更大的纤维素转化率,但生成葡萄糖反应的选择性较低,副产物较多,并且稀酸催化水解需要在较高的温度下进行。稀酸水解工艺包括 Scholler 工艺、Madison 工艺、Grethlein 工艺、两段法工艺等。20 世纪 20 年代发展起来的 Scholler 工艺是第一种纤维素酸水解技术。在这个工艺中,0.5%(质量分数)的硫酸溶液在 170℃、2MPa 的条件通入渗滤器,并与渗滤器中的木材废料接触约 45min。最后,将水解所得的稀糖溶液从渗滤器中移出,冷却,中和,得到可发酵的糖溶液。该过程中还原糖的收率达到 50%。在 Madison 工艺中,木材在 150~180℃ 的温度下用 0.5%(质量分数)H_2SO_4 连续流动处理。这种连续工艺比 Scholler 工艺效率高得多,因为水解产物高温下在反应器中停留的时间很短。

酸水解纤维素副反应较多,如何避免产物葡萄糖继续转化生成 5-羟甲基糠醛、乙酰丙酸等物质仍然需要进一步研究,且不论是稀酸还是浓酸水解纤维素,都要求设备耐腐蚀性好。实现酸水解纤维素工业化所投入的大量酸对环境污染大,废液处理成本高。

(2)均相酸性金属盐催化剂

一些能溶于水的酸性金属盐[如 $CuCl_2$、$AlCl_3$、$CrCl_3$、$SnCl_2$、$ZnCl_2$、$Al_2(SO_4)_3$、$Fe_2(SO_4)_3$ 等]也被用于纤维素的水解。这些金属盐对纤维素水解的催化作用相对弱于无机酸,所以通常需要在较高的温度下(通常超过 180℃)才能较好地催化纤维素水解。但是它们通常对葡萄糖的后续转化具有特殊的选择性,故它们常用于催化纤维素转化直接制取 HMF、乙酰丙酸(酯)、乳酸(酯)等平台化学品。

(3)非均相催化剂

非均相催化剂的使用可以克服均相催化剂使用过程中所存在的反应器腐蚀、催化剂回收难等诸多问题,所以部分研究尝试用非均相催化剂去催化纤维素水解。然而,由于固体催化剂与纤维素之间存在固-固传质阻碍,使得纤维素表面能够与固体催化剂活性位点相接触的糖苷键较少,所以固体酸对纤维素水解的催化作用较弱。目前报道较多的对纤维素具有较好催化作用的固体催化剂主要包括磺化碳材料、酸性树脂、复合金属氧化物。

磺化碳材料是一类研究较多的纤维素水解固体催化剂。Suganuma 等研究了一系列固体催化剂对纤维素水解为葡萄糖的催化作用,发现传统的固体酸,如铌酸、H-丝光沸石、Nafion 和 Amberlyst-15,都不能有效地催化纤维素的水解,而含有磺酸基、羧基和羟基官能团的无定形碳材料对纤维素的水解具有较好的催化作用[21]。他们认为,磺化碳催化剂的催化性能归因于碳材料中的羧基、羰基和羟基对纤维素中的 β-1,4-葡聚糖的吸附能力,从而促进磺酸基官能团催化被吸附的 β-1,4-糖苷键发生水解断裂(图 2-2)。磺化碳材料不仅对纤维素水解具有较好的催化作用,而且该类材料在水热条件下较为稳定,因此,后续大量工作继续深入研究了将各种磺化碳材料用于催化纤维素水解[22-25]。Pang 等采用在 250℃ 磺化的 CMK-3 催化纤维素水解可得到 74.5% 的葡萄糖[23]。Van De Vyver 等合成了一种磺化碳硅复合材料 Si-C-SO$_3$H,并发现它能够有效促进纤维素水解,他们认为这些磺化碳硅复合固体酸中的 Si-C 表面能够对纤维素的糖苷键产生吸附作用[26]。由于生物质在水热条件下可以发生碳化得到含有羰基、羟基和羧基的碳材料,因此,部分研究尝试将生物质水热碳化得到的水热焦炭用于制备纤维素水解的催化剂。Guo 等采用葡萄糖不完全水热碳化再磺化的方法制备了磺化碳材料。该碳材料含有磺酸基、羧基和羟基官能团,能够在离子液体 1-丁基-3-甲基咪唑氯化物中催化纤维素水解,得到的总还原糖收率为 72.7%[27]。

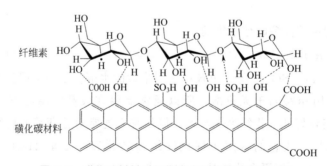

图 2-2 磺化碳材料对纤维素水解的催化作用机理

磺化树脂如 Amberlyst-15、Amberlyst-35、Nafion-NR50 和 Nafion-SAC-13 等是一类常用的酸性催化剂,但是它们对纤维素水解反应的催化作用较弱[28]。这可能是由于这些催化剂的活性位点与纤维素的糖苷键相互接触的机会较少。Shuai 等合成了一种磺化氯甲基聚苯乙烯树脂 CP-SO$_3$H,发现该催化剂能够非常有效地催化纤维素和纤维二糖的水解。他们发现这种新型的CP-SO$_3$H中的氯原子作为氢键受体能够与纤维素形成较强的吸附，而催化剂中的磺酸基官能团则作为酸性位点催化纤维素中糖苷键的断裂，两种活性位点的协同作用降低了纤维素水解的活化能（图2-3）[29]。不过，这些酸性树脂在高温水热条件下的稳定性仍然需要进一步研究。

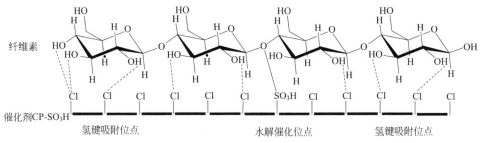

图 2-3　催化剂 CP-SO$_3$H 与纤维素之间的相互作用示意图

酸性复合氧化物，如氧化钽、磷酸钽、磷酸锆、磺酸化的二氧化锆、钨酸铝、钨酸锆等，是另一类纤维素催化水解中研究较多的固体酸[18,30-33]。其中，钨基固体酸如钨酸铝、钨酸锆是一类水热稳定性较好且对纤维素水解具有催化作用的固体催化剂。Chambon等发现路易斯酸AlW和ZrW均能够较有效地催化纤维素水解[30,34]。Hamdy采用共沉淀的方法利用钨酸钠和硫酸铝制备了钨酸铝催化剂，并发现该催化剂对纤维素的水解具有较好的催化作用[35]。通过分析对纤维素水解具有催化作用的固体催化剂的性质，发现所有对纤维素具有较好催化作用的固体催化剂结构中都存在氢键受体，如磺化碳材料中的C=O物种，钨酸铝中的W=O物种，以及磺化氯甲基聚苯乙烯树脂中的C=O物种和C—Cl物种。这些研究表明，固体催化剂与纤维素表面的羟基形成氢键可能对纤维素水解具有重要作用。因此，作者认为，在钨酸铝催化纤维素水解时，钨酸铝的W=O物种可能和纤维素表面的羟基形成氢键和配位键，使得纤维素和钨酸铝催化剂能够形成较强的相互作用，而钨酸铝中的Al—OH物种则作为活性位点（布朗斯特酸中心）催化纤维素的糖苷键断裂（图2-4）。

因为木质纤维素在水解后剩余的木质素会与固体催化剂混合在一起而难以分离，所以，部分研究将具有磁性的Fe$_3$O$_4$引入催化剂中制备出具有磁性的固体催化剂，在反应结束后可以通过永磁体将催化剂吸附，从而实现催化剂与水解残渣分离[36-38]。Zhang等将磁性核心Fe$_3$O$_4$包裹在磺化碳壳中制备出具有核壳结构的Fe$_3$O$_4$@C-SO$_3$H纳米催化剂，该催化剂的磺化碳壳能够有效地

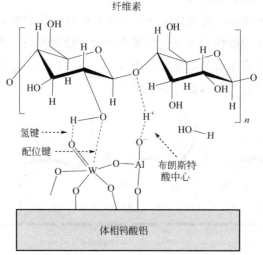

图 2-4　钨酸铝的活性位点及它对纤维素水解的催化机理

催化纤维素水解为葡萄糖，而磁性核心 Fe_3O_4 使得催化剂很容易通过外加磁场从反应混合物中分离出来，从而解决了催化剂与反应后剩余木质素残渣的分离问题。

2.3　催化碳水化合物转化制备 5-羟甲基糠醛的原理

2.3.1　5-羟甲基糠醛简介

5-羟甲基糠醛（HMF）是己糖脱去三个水分子后生成的一种呋喃杂环化合物，可以用于生产多种生物质衍生化学品和高品质液体燃料，因而受到国内外研究者的广泛关注[39-41]。如图 2-5 所示，HMF 可以经过水合反应制取乙酰丙酸[42]，与低碳醇发生醚化反应生成 5-烷氧基甲基糠醛[43-45]，经加氢脱氧反应生成 2,5-二甲基呋喃[46]，或者经选择性氧化生成 2,5-呋喃二甲醛和 2,5-呋喃二甲酸[47]。其中，2,5-二甲基呋喃的热值与汽油相当，可作为燃料添加剂掺入汽油中使用[46]；2,5-呋喃二甲醛可作为医药中间体、杀菌剂、交联剂、大环配体的中间体以及功能性材料单体[48]；而 2,5-呋喃二甲酸被认为是一种在聚对苯二甲酸乙二醇酯（PET）生产中可能替代对苯二甲酸的单体[41,49]。此外，HMF 可以与其他羰基化合物发生羟醛缩合反应得到长链含氧化合物，然后经由加氢脱氧反应得到长链烷烃[50-55]。

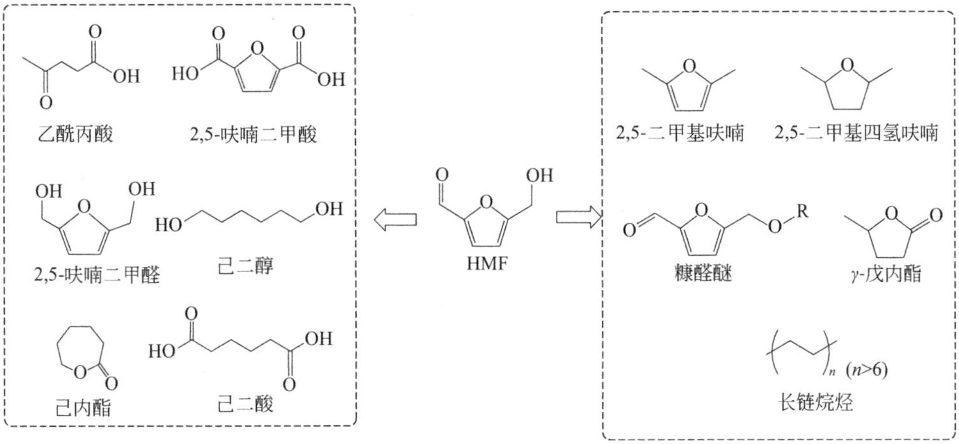

图 2-5　由 HMF 制得的化学品及液体燃料

由于 HMF 在制备生物质基液体燃料和化学品方面所展示出的巨大潜力，近年来许多研究者在碳水化合物制备 HMF 的反应路径、反应机理、反应介质和催化剂体系等方面进行了深入研究，取得了引人注目的进展[39,40,56-63]。

2.3.2　己糖制备 5-羟甲基糠醛的反应机理

葡萄糖和果糖是两种自然界中最常见的己糖，它们在酸性催化剂的作用下脱去三分子水即可生成 HMF。其中，果糖比葡萄糖更容易转化为 HMF，但是果糖的价格比葡萄糖昂贵，因此，以葡萄糖为原料制备 HMF 比利用果糖制备 HMF 更具研究价值。

图 2-6 展示了葡萄糖脱水生成 HMF 的路径及所涉及的主副反应。可以看到，葡萄糖可通过两种路径生成 HMF：①葡萄糖发生酮式-烯醇互变异构反应生成果糖，果糖经缩醛环化后脱去三分子水而生成 HMF；②葡萄糖通过 β-消除和酮式-烯醇互变异构反应生成 3-脱氧葡萄糖醛酮（3-DG），而 3-DG 再经缩醛环化和脱水反应生成 HMF[64]。

通常认为，果糖比葡萄糖更容易生成 HMF 的主要原因是：葡萄糖在水溶液中几乎完全以吡喃环（六元环）形式存在[65]，所以葡萄糖在生成 HMF 的过程中首先需要开环形成链式葡萄糖，然后链式葡萄糖再经异构化生成果糖；相反，果糖在溶液中有 21.5% 的呋喃环式异构体存在[66]，使得果糖在脱水生成 HMF 的过程中不需要经历开环过程。然而，笔者研究发现，α-羰基醛是碳水化合物生成水热焦炭的关键前驱体，而葡萄糖在水热条件下比果糖更容易生成 α-羰基醛，因此，笔者认为葡萄糖生成 HMF 的选择性比果糖低的另一个重要原因是葡萄糖比果糖更容易生成水热焦炭等副产物[67,68]（第五章中详细论述）。如图2-6所示，葡萄糖是醛糖，它在水热条件下易通过β-消除反应和酮式-烯醇互变反应而生成3-DG；另一方面，HMF在水热条件下易发生水解开环而生成2,5-二氧代-6-羟基己醛和2,5-二氧代-3-己烯醛，这两种物质同样是α-羰基醛。所以，葡萄糖在水热降解过程中易生成3-DG、2,5-二氧代-6-羟基己醛和2,5-二氧代-3-己烯醛这几种α-羰基醛，它们容易发生羟醛缩合反应而生成水热焦炭，从而降低葡萄糖生成HMF的选择性；相反，果糖是酮糖，它在水热条件下不易生成α-羰基醛，因此果糖在水热条件下生成HMF的选择性比葡萄糖高。

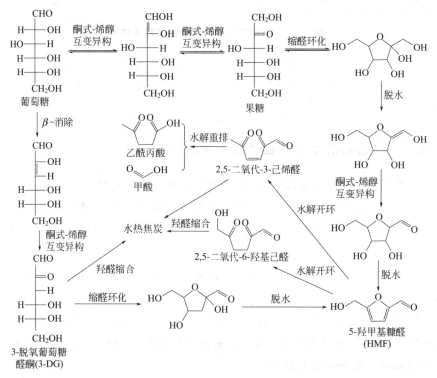

图2-6 葡萄糖脱水生成HMF的路径及所涉及的主要副反应[59,64,69]

2.3.3 反应介质、催化剂及反应原料

反应介质和催化剂是影响碳水化合物转化为HMF效率最重要的两个参数，对利用生物质原料制备HMF技术的经济性具有重要影响。目前人们开发了多种反应介质和催化剂以提高碳水化合物转化为HMF的效率。

2.3.3.1 反应介质

（1）水

水是转化木质纤维素生产各类平台化学品最理想的反应溶剂。然而，在水中转化葡萄糖和果糖制备HMF的收率往往难以超过30%。这是因为在酸性水溶液中HMF极易发生水解开环生成乙酰丙酸、甲酸和水热焦炭。Asghari等在270℃的高温水溶液中用磷酸转化纤维素能够得到约25%的HMF[70]。笔者曾尝试以水蒸气作为反应介质，在固定床中采用硫酸氢盐、磷酸二氢盐等酸式盐催化纤维素转化制取HMF，但是收率也难以超过30%[71,72]。未来的研究中可尝试采用气体作为反应介质，或者在真空条件下去转化碳水化合物制取HMF。

（2）单相极性有机溶剂体系

由于采用水作为溶剂转化碳水化合物制取HMF时，HMF易发生水解开环反应而生成乙酰丙酸和水热焦炭，因此，部分研究尝试采用极性非质子有机溶剂如 N-甲基吡咯烷酮（NMP）、N,N-二甲基甲酰胺（DMF）、N,N-二甲基乙酰胺（DMA）、N,N-二甲基亚砜（DMSO）、己内酰胺（CPL）等作为反应溶剂去转化碳水化合物制取HMF。在这些极性有机溶剂转化碳水化合物时，由于HMF的水解开环反应受到抑制，因此HMF收率可以得到明显的提高。以果糖为原料时，在上述各类有机溶剂中可以得到60%~100%的HMF[73]，特别是以DMSO为溶剂时能够非常容易地由果糖脱水得到超过70%的HMF。此外，将碱金属盐类加入有机溶剂中形成的催化体系对葡萄糖制取HMF具有独特催化作用。Binder等在由DMA与卤化锂盐（LiCl、LiBr）组成的体系中，采用卤化铬盐（$CrCl_2$、$CrCl_3$）催化葡萄糖的转化，得到了70%以上的HMF[74]。

在极性非质子有机溶剂中转化碳水化合物制取HMF具有条件温和、HMF收率高、反应时间短等优点。但是，利用这些有机溶剂作为反应介质用于制备HMF存在诸多难以克服的问题，比如：①这些有机溶剂价格昂贵，导致大规模制备HMF所需的成本高昂；②这类反应溶剂的沸点通常较高，导致将产物HMF从反应溶剂中通过蒸馏分离出来所需的能耗高；③这些极性有机溶剂对环境有污染，不能直接排放到环境中，需在反应结束后进行回收重复利用，但这又同样增加了生产成本。这些缺陷阻碍了这类溶剂在大规模工业化制备HMF上的应用。

（3）离子液体

离子液体是指在室温下呈液态的自由离子构成的液体。离子液体中，阴离子电负性很强,同时阳离子半径较大,这样就能有效削弱阴阳离子间的库仑力,从而降低熔点,使得物质在较低的温度下熔化,呈液体状。咪唑类离子液体,比如[HMIM]Cl、[EMIM]Cl、[BMIM]Cl、[EMIM]BF$_4$、[OMIM]Cl等,常被用作转化碳水化合物制备HMF的反应介质[75]。当离子液体作为碳水化合物转化的反应介质时,由于离子液体间存在离子键,有很强的静电场,能够改变分子内的键能[76],使得葡萄糖和果糖的脱水反应能够在较温和的反应条件下进行(通常反应温度在80~120℃),且得到较高的HMF收率。比如,Zhao等在离子液体[EMIM]Cl中采用CrCl$_2$催化葡萄糖降解,在100℃的温和条件下,可以得到69%的HMF收率[77]。

虽然离子液体作为反应介质制备HMF具有选择性高、反应条件温和、反应速率快和副反应少等优点,但由于离子液体本身制备工艺复杂、造价昂贵、回收困难,同样不宜用作大规模转化木质纤维素的反应溶剂。

(4)水-有机溶剂组成的双相反应介质

如前所述,HMF的水解开环反应会导致HMF在酸性水溶液中不稳定[78]。因此,部分研究者开发了由水和与水不互溶的有机溶剂所组成的双相反应介质[61,79-81]。在双相反应体系中,碳水化合物和催化剂主要分散于水相中,而反应生成的HMF则被萃取到有机相中,从而减少HMF的水解开环反应。双相反应体系中己糖脱水制备HMF的示意图见图2-7。

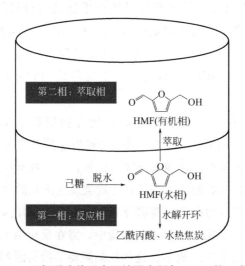

图2-7 双相反应体系中己糖脱水制备HMF的示意图

有机溶剂的选择是影响双相体系性能的一个关键因素。通常需要所选用的有机溶剂与水不互溶,对HMF具有较好的萃取作用,且沸点较低,易于从有机溶剂中分离出 HMF。目前研究较多的双相体系采用的有机溶剂包括 1-丁醇、2-丁醇、四氢呋喃、γ-戊内酯、甲基四氢呋喃、甲基异丁基酮（MIBK）、烷基酚等[46,61,82]。上述溶剂中，四氢呋喃、γ-戊内酯和烷基酚都可以由木质纤维素转化得到，从而减少对化石资源的依赖[66]。

2.3.3.2 催化剂

由于己糖生成 HMF 的过程涉及脱水反应，故研究大都采用对脱水反应具有催化作用的酸性催化剂来催化这一过程。研究较多的用于制备 HMF 的催化剂可以分为液体酸、酸性金属盐、固体酸或上述多种催化剂的组合。表 2-1 总结了用于己糖脱水制备 HMF 的催化剂[73]。其中，在离子液体和极性非质子有机溶剂中，无机酸和金属盐具有较好的催化效果，而在以水为反应介质的体系中，固体酸是研究较多的催化剂。

表 2-1 用于己糖脱水制备 HMF 的催化剂

催化剂类别		催化剂类别
液体酸	有机酸	甲酸、乙酸、草酸、乙酰丙酸、马来酸、对甲苯磺酸等
	无机酸	磷酸、硫酸、盐酸、氢碘酸等
酸性金属盐		$CrCl_2$、$CrCl_3$、$SnCl_4$、$AlCl_3$、$Al_2(SO_4)_3$、$ZnCl_2$、$ZnSO_4$、$GeCl_4$、$ScCl_3$、$YbCl_3$ 等
固体酸		离子交换树脂（Amberlyst-15 等）、分子筛（HY、HMOR、H-ZSM-5、Hβ、SAPO-34、SBA-15-SO_3H、Fe_3O_4@SiO_2-SO_3H 等）、金属氧化物（TiO_2、ZrO_2、SO_4^{2-}/ZrO_2、Nb_2O_5、$NbOPO_4$、Ta_2O_5 等）、磺化碳材料、杂多酸（$H_3PW_{12}O_{40}$/MIL-101、$FePW_{12}O_{40}$、$Ag_3PW_{12}O_{40}$、$Cs_{2.5}H_{0.5}PW_{12}O_{40}$）
组合催化剂		$NaHSO_4$ + $ZnSO_4$、Sn-β + HCl、Nb_2O_5 + HCl、$AlCl_3$ + HCl、$ZnCl_2$ + HCl、$CrCl_3$ + HCl

2.3.3.3 反应原料

己糖（葡萄糖和果糖）及能够水解生成己糖的物质（蔗糖、菊芋粉、淀粉、纤维素、木质纤维素）均可作为生产 HMF 的原料。然而，因为原料的性质不同，制备 HMF 的难易程度相差很大。果糖是酮糖，转化过程中发生的副反应较少，因此果糖是最易于制备 HMF 的原料；但果糖在自然界中存在量极少，价格较为昂贵，不适于大规模用作生产 HMF 的原料；木质纤维素在自然界中储量巨大，是最适于用作大规模生产 HMF 的原料，但由于木质纤维素结构复杂且其中的纤维素难水解，而生成的 HMF 稳定性又较差，导致采用木质纤维素制备 HMF 的难度较大，选择性较低。

（1）单糖

酸催化果糖脱水是制备 HMF 最方便的方法。大量研究采用液体无机酸

（HCl、HBr、H_2SO_4）为催化剂，在离子液体或双相体系中转化果糖都能够得到70%以上的HMF[46,82-88]。然而，以果糖为原料大规模工业制备HMF的成本较高。

与果糖不同，葡萄糖可以通过水解淀粉、纤维素或含有纤维素的生物质得到，所以其来源更为广泛。然而，利用葡萄糖制备HMF的选择性通常都较低。2007年，Zhao等在[EMIM]Cl中使用$CrCl_2$催化葡萄糖脱水，首次得到高达70%的HMF收率。他们提出，在离子液体中，Cr^{2+}对葡萄糖经历烯醇式结构转化为果糖具有很好的催化效果，这是$CrCl_2$能够高效催化葡萄糖制备HMF的原因[77]。继Zhao等的研究之后，大量的在各类离子液体中用铬盐催化葡萄糖基糖类制备HMF的研究工作被陆续报道。Yong等在[BMIM]Cl中分别使用$CrCl_2$和$CrCl_3$与N-杂环碳烯催化葡萄糖转化制备HMF，均可以得到81%的HMF收率[89]。由于Cr^{3+}对葡萄糖异构为果糖有催化作用，含铬的各种物质是研究最多的一类催化剂[77,90-93]。其他的一些金属盐，比如$ZnCl_2$、$GeCl_4$、$SnCl_4$、$AlCl_3$等也被报道对催化葡萄糖转化为HMF具有较好的催化效果[94-98]。Yang等在水-四氢呋喃（THF）双相体系中采用$AlCl_3$转化葡萄糖，得到的HMF收率可达到52%[97]。此外，部分研究也将一些路易斯酸和布朗斯特酸结合去催化葡萄糖转化制备HMF。比如$AlCl_3$+HCl、$ZnCl_2$+HCl、$CrCl_3$+HCl、Sn-β+HCl等催化剂体系都能够有效地催化葡萄糖制备HMF[66,99-101]。

虽然采用金属盐做催化剂催化葡萄糖脱水能够得到很高的HMF收率，但由于金属盐催化剂与产物分离困难，且大部分金属盐有毒（$CrCl_2$、$CrCl_3$、$SnCl_4$、$GeCl_4$均有剧毒），致使后续处理工序复杂，且造成严重的环境污染。相比于液体酸和金属盐催化剂，固体催化剂具有与产物分离容易、环境友好、能重复回收利用等一系列优点。近几年，有学者对固体酸催化葡萄糖和果糖脱水进行了大量研究工作。研究中常用到的固体酸有ZrO_2、TiO_2、SO_4^{2-}/ZrO_2、SO_4^{2-}/ZrO_2-Al_2O_3、Sn-β、Nb_2O_5、H-ZSM-5、$Ag_3PW_{12}O_{40}$及磺化碳材料等[102-110]。在上述固体催化剂中，Sn-β分子筛对醛糖异构为酮糖具有较好的催化效果，是一种极有应用潜力的固体酸[111,112]。Nikolla等将Sn-β与HCl结合协同催化葡萄糖制取HMF，在水/THF中可得到56.9%的HMF收率[112]。

（2）纤维素

直接采用纤维素及木质纤维素制备HMF更具有原料来源广泛的优势，从而更具有工业应用前景。然而，由于纤维素结构较为稳定而 HMF 的稳定性又较差[113-115]，使得利用纤维素制备 HMF 成为一个巨大的挑战[116-119]。表 2-2 列出了一些采用纤维素制备 HMF 的代表性研究成果。

表 2-2　采用纤维素制备 HMF 的代表性研究成果

反应介质	催化剂	反应条件	收率/%	参考文献
水	无	280℃,4min	12	[120]
水	磷酸	270℃,2min	30	[70]
水	α-Sr(PO$_3$)$_2$	230℃,5min	15	[121]
[EMIM]Cl	无	120℃,3h	21	[125]
[BMIM]Cl	GeCl$_4$	120℃,30min	35	[133]
[BMIM]Cl	HY/Ipr-CrCl$_2$	120℃,12h	48	[134]
[EMIM]Cl	CuCl$_2$+CrCl$_2$	120℃,8h	55	[126]
[BMIM]Cl	CrCl$_3$	400W 微波加热 2min	62	[127]
[BMIM]Cl	CrCl$_3$	150℃,10min	55	[135]
[BMIM]Cl	CrCl$_3$+LiCl	160℃,10min	62	[93]
[EMIM][Ac]	[C$_4$SO$_3$-HMIM][CH$_3$SO$_3$]+CuCl$_2$	160℃,3.5h	69.7	[128]
DMA-LiCl	HCl+CrCl$_3$	140℃,2h	54	[74]
H$_2$O-MIBK	TiO$_2$	270℃,1h	35	[136]
H$_2$O-THF	AlCl$_3$	180℃,0.5h	37	[97]
H$_2$O-THF	NaHSO$_4$+ZnSO$_4$	160℃,1h	53	[20]
H$_2$O-GVL	Al$_2$(SO$_4$)$_3$	165℃,50min	43.5	[132]

以水作为反应介质转化纤维素通常仅能得到较低的 HMF 收率。Ehara 等在 280℃亚临界水中不使用催化剂转化纤维素，仅得到 11.9% 的 HMF[120]。Asghari 等在 270℃ 的水热条件下用磷酸转化纤维素，得到约 25% 的 HMF[70]。Daorattanachai 等在 230℃热水中以固体酸 α-Sr(PO$_3$)$_2$ 作为催化剂转化纤维素得到 15% 的 HMF[121]。

离子液体是优良的氢键受体，甚至能够在室温下有效地溶解纤维素[122,123]。因此，利用离子液体作为溶剂转化纤维素制备 HMF 可以在更温和的反应条件下进行[124,125]。大多数以离子液体作溶剂的反应体系中常用氯盐作为催化剂，比如 CrCl$_3$、CuCl$_2$+CrCl$_2$、CrCl$_2$+RuCl$_3$ 等，得到 HMF 收率在 50%～70% 之间[93,126-129]。Ding 等在离子液体 [EMIM][Ac] 中，采用 CuCl$_2$ 和 [C$_4$SO$_3$-HMIM][CH$_3$SO$_3$] 作为催化剂转化纤维素，得到 69.7% 的 HMF[128]。Zhang 等采用 CuCl$_2$ 和 CrCl$_2$ 组成的催化剂在 [EMIM]Cl 中转化纤维得到55%的HMF[129]。

除了离子液体，极性非质子有机溶剂（DMSO、DMF、DMA等）由于能够抑制HMF的水解开环反应，也常被用作制备HMF的反应介质[130,131]。Binder等采用DMA-LiCl作为反应溶剂，在HCl和$CrCl_3$的催化作用下，转化纤维素可以得到54%的HMF[74]。

由水与有机溶剂组成的双相体系在转化纤维素制备HMF方面也表现出优异的性能。Yang等在水与四氢呋喃组成的双相体系中利用$AlCl_3$催化纤维素转化得到37%的HMF[97]。笔者在水与四氢呋喃组成的双相体系中利用$NaHSO_4$+$ZnSO_4$协同催化纤维素转化得到53%的HMF[20]。Qi等采用$Al_2(SO_4)_3$催化剂经球磨后处理纤维素，在由水和γ-戊内酯（GVL）组成的双相体系中转化可得到的HMF收率达到43.5%[132]。需要说明的是，一些从生物质转化中得到的有机溶剂，如烷基酚、γ-戊内酯等，被证明能够用作双相体系中的萃取相用于制备HMF[66,101,132]，这使得在双相体系中制备HMF有望实现工业应用。

2.4 木质纤维素制备乙酰丙酸及其衍生酯类的原理

2.4.1 乙酰丙酸简介

乙酰丙酸（levulinic acid）是己糖在水热条件下生成的另一种重要的平台化学品。如图2-8所示，乙酰丙酸可通过多种反应生成大量的下游化学品，如琥珀酸、丙烯酸、戊酸、5-氨基乙酰丙酸、乙酰丙酸酯、乙酰丙酸钠、α-当归内酯、γ-戊内酯、1,4-丁二醇、1,4-戊二醇、2-甲基四氢呋喃、THF等，而这些化学品又可作为原料生产树脂、除草剂、医药、助燃剂、溶剂、增塑剂、防冻剂、液体燃料和含氧燃料添加剂等[137,138]。

在多种乙酰丙酸衍生化学品中，乙酰丙酸酯、2-甲基四氢呋喃和γ-戊内酯可以很容易地与石油产品混合，因此它们可作为液体燃料添加剂加入液体燃料中。特别地，2-甲基四氢呋喃是一种具有高燃烧性能的燃料添加剂，当添加量达到30%时，仍然不降低汽油的燃烧性能且不需要对发动机进行改装[56]。另外，乙酰丙酸转化得到的戊酸、α-当归内酯、γ-戊内酯均可通过后续的碳链增长、加氢脱氧而转化为碳链长度大于7的液态烷烃，而这些液态烷烃可以直接作为液体燃料使用[139,140]。

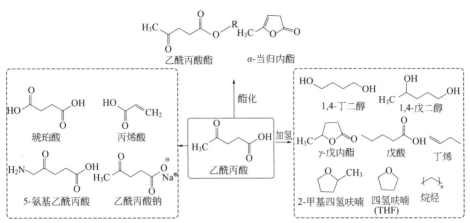

图 2-8　乙酰丙酸作为平台化学品生产化工产品和液体燃料[141-143]

2.4.2　己糖生成乙酰丙酸的原理

通常认为，由己糖转化成乙酰丙酸需要经过以下两个主要步骤：①己糖经脱水形成 HMF；②HMF 在酸催化剂的作用下水解成甲酸和乙酰丙酸。关于 HMF 在酸性介质中脱水生成乙酰丙酸的反应机理，Horvat 等[144] 利用核磁技术证明 2,5-二氧代-3-己烯醛是 HMF 生成乙酰丙酸的关键中间体。他们指出，HMF 在酸催化剂的作用下发生水解、脱水、开环等一系列反应并生成 2,5-二氧代-3-己烯醛；随后，2,5-二氧代-3-己烯醛发生 C—C 键水解断裂而释放出甲酸并生成 4-氧代-2-戊烯醛；最后，4-氧代-2-戊烯醛通过水合、重排反应生成乙酰丙酸（图 2-9）。

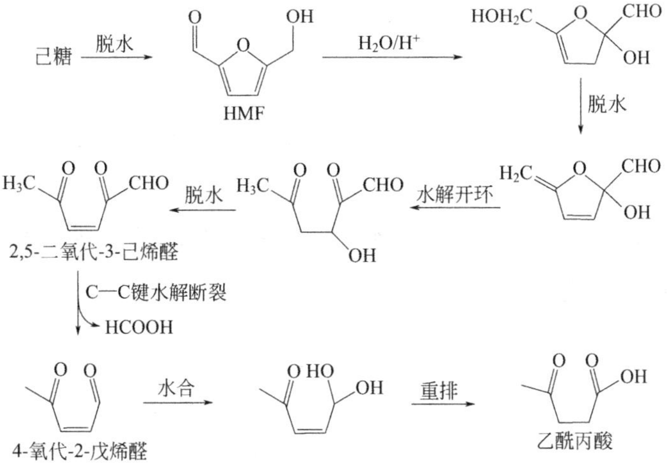

图 2-9　Horvat 提出的己糖生成乙酰丙酸的路径[145]

葡萄糖生成乙酰丙酸的选择性并不比果糖低[146]，而葡萄糖生成 HMF 的选择性明显低于果糖生成 HMF 的选择性，表明葡萄糖可能存在不经 HMF 而通过其他中间体生成乙酰丙酸的路径。如前所述，在葡萄糖经 HMF 生成乙酰丙酸的路径中，HMF 水解开环所生成的 2,5-二氧代-3-己烯醛是一种 α-羰基醛，而它发生 C—C 键水解断裂反应是生成乙酰丙酸的关键步骤。大量研究表明，α-羰基醛在水热条件下易发生 C—C 键水解断裂反应[147-149]，而 3-脱氧葡萄糖醛酮是葡萄糖脱水生成的一种重要 α-羰基醛，因此，笔者认为，葡萄糖在水热条件下可通过 3-脱氧葡萄糖醛酮生成乙酰丙酸。如图 2-10 所示，笔者认为葡萄糖经 β-消除和酮式-烯醇互变异构生成的 3-脱氧葡萄糖醛酮可以发生 C—C 键水解断裂而生成 2-脱氧核糖，而 2-脱氧核糖则能够脱水生成糠醇，并进而水解生成乙酰丙酸。因此笔者认为葡萄糖生成乙酰丙酸的路径有两条，分别是以 HMF 和以 3-脱氧葡萄糖醛酮为平台。不过，目前葡萄糖以 3-脱氧葡萄糖醛酮为中间体而生成乙酰丙酸的路径还没有得到实验证据。

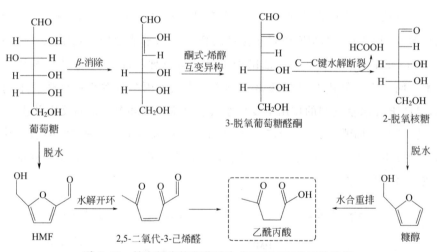

图 2-10 笔者提出的葡萄糖生成乙酰丙酸的两种路径

2.4.3 木质纤维素及其衍生碳水化合物生成乙酰丙酸的催化剂及生产工艺

相比于生成 HMF 而言，利用纤维素乃至生物质制备乙酰丙酸的技术要容易得多，这是因为乙酰丙酸在水中的稳定性较高。无机强酸如硫酸和盐酸经常被用于催化己糖水解生产乙酰丙酸[42,150-153]。早在 1990—1997 年，美国 Biofine 公司开发了一种以硫酸催化木质纤维素转化制备乙酰丙酸的生产工艺[138,154,155]。

如图2-11所示，在Biofine工艺中，木质纤维素与硫酸[1.5%～3%（质量分数）]混合，在210～230℃的条件下进入第一个反应器反应13～25s，生成以HMF和单糖为主的降解液；随后，在第一个反应器中得到的反应液不断进入第二个反应器，在190～220℃的温度下继续反应15～30min而得到乙酰丙酸。利用Biofine工艺生产的乙酰丙酸收率高达70%～80%，高于其他工艺所生产的乙酰丙酸收率。然而，Biofine工艺生产乙酰丙酸及衍生物的技术在商业化生产上面临着严峻的挑战。第一，乙酰丙酸需要从无机酸催化剂中分离出来，以循环使用酸催化剂，避免在下游工艺中产生负面影响。第二，乙酰丙酸的生产浓度低，提纯/回收过程成本较高。第三，该技术中所产生的水热焦炭易堵塞管道。此外，该过程中副产物固体残渣作为燃料在进行燃烧之前需要去除其吸附的无机酸催化剂，这又导致新的污染。

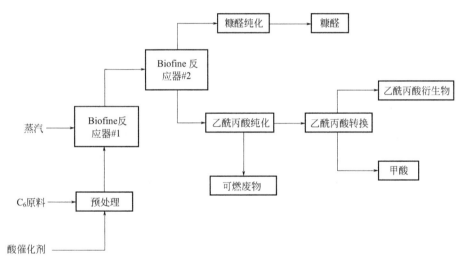

图2-11 生产乙酰丙酸的Biofine工艺[138]

由于大量的文献都认为葡萄糖生成乙酰丙酸的过程涉及葡萄糖异构化为果糖和果糖脱水生成HMF这两个关键步骤，而路易斯酸对催化葡萄糖异构化为果糖具有较好的催化作用[77]，因此一些路易斯酸，如$CrCl_3$、$FeCl_3$、$CuCl_2$、$AlCl_3$、$CuSO_4$和$Fe_2(SO_4)_3$等，也都被用于催化木质纤维素转化为乙酰丙酸，并发现具有较好的催化作用[19,156-158]。特别地，Peng等采用$CrCl_3$在水热条件下催化纤维素降解能够得到67%的乙酰丙酸[19]。此外，部分研究尝试采用路易斯酸和布朗斯特酸协同催化葡萄糖转化为乙酰丙酸。如$CrCl_3$+HCl、$CrCl_3$+H_3PO_4等二元催化剂体系都被证明对转化葡萄糖制备乙酰丙酸具有较好的催化作用[158,159]。这些金属盐的腐蚀性相对较低，比使用液体无机酸更加安全。不过，从分离回收、毒性的角度考虑，这些金属盐并没有显著地优于液体无机酸。

为了解决均相催化剂使用中存在的催化剂分离回收的难题，部分研究报道了采用固体酸催化剂转化木质纤维素制取乙酰丙酸。例如，分子筛[37,160]、金属氧化物[161,162]、酸性（磺化或磷酸化）金属氧化物[161,163,164]、酸性树脂[165-170]、磺化碳材料[171]等都被报道对转化木质纤维素及其衍生物己糖制备乙酰丙酸具有较好的催化作用。Zuo 等[165]使用磺化的氯甲基聚苯乙烯作为固体催化剂转化纤维素制取了收率高达 65% 的乙酰丙酸。

值得一提的是，纤维素除了可以通过水解为葡萄糖，再由葡萄糖转化乙酰丙酸这一条路径，也在水相中直接部分氧化纤维素制备乙酰丙酸。Lin 等使用 ZrO_2 作为催化剂，在 240℃ 下，以 3.5MPa 的 N_2-O_2 混合气体作为氧化剂去氧化纤维素得到收率为 50% 的乙酰丙酸。他们认为，该过程中纤维素首先转化为葡萄糖酸，葡萄糖酸再经脱羧转化为 2-脱氧核糖，而 2-脱氧核糖则脱水生成糠醇并进而转化为乙酰丙酸[172,173]。该方法的优势在于，反应过程中只产生极少量的水热焦炭，可以避免催化剂被水热焦炭包裹而失活，同时也避免反应过程中生成的大量水热焦炭堵塞反应器。

乙酰丙酸的分离和回收也是利用木质纤维素制备乙酰丙酸工艺的关键技术。由于反应过程中原料浓度不高，所以在生物质降解后得到的降解液中乙酸乙酯的浓度较低，而将这些低浓度的乙酸乙酯分离出来需要消耗大量的能量。Kang 等将木质纤维素转化后得到的含有乙酸乙酯的降解液作为循环溶液去处理新的生物质原料，经过反应液的多次循环使用，可得到高浓度的乙酰丙酸（105g/L）和甲酸（39g/L），降低分离成本[174]。

2.4.4 利用木质纤维素制备乙酰丙酸酯

乙酰丙酸酯是一类广泛应用于食品工业的化合物，常用作溶剂和增塑剂。此外，它还表现出一些特性，如毒性低、润滑性高、闪点高、稳定性好和低温条件下的中等流动性能，使它适合用作生物柴油中的低温流动性改进剂，以及汽油和柴油中的含氧添加剂[175-177]。部分研究表明，乙酰丙酸酯可以与柴油混合，含有 20% 乙酰丙酸酯、79% 柴油和 1% 其他助剂的混合物是一种柴油发动机中硫排放较低的燃料[178]。

如图 2-12 所示，利用木质纤维素生产乙酰丙酸酯的路径有三种[179,180]：①木质纤维素中的纤维素通过酸催化水解得到 HMF，而 HMF 发生水合生成乙酰丙酸，随后乙酰丙酸与低碳醇进行酯化反应得到乙酰丙酸酯[181]；②木质纤维素中的半纤维素通过酸催化汽提工艺得到糠醛，糠醛经加氢得到糠醇，糠醇再在醇溶剂中发生醇解得到乙酰丙酸酯[182]；③木质纤维素中的纤维素及其衍生碳水化合物直接醇解制备乙酰丙酸酯，即在醇溶剂中在酸催化下纤维素醇解生成烷基葡萄糖苷，而烷基葡萄糖苷经加热进一步脱水生成5-烷氧基甲基糠醛，随后5-烷氧基甲基糠醛进一步醇解生成等物质的量的乙酰丙酸酯和甲酸酯[183]。

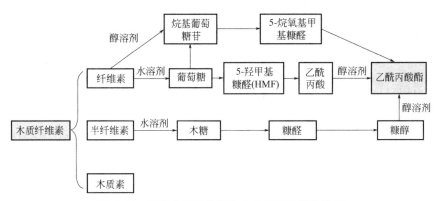

图 2-12　利用木质纤维素生产乙酰丙酸酯的路径

在上述三种利用木质纤维素制备乙酰丙酸酯的路径中，通过生物质直接醇解制备乙酰丙酸酯的路径具有诸多优点：①该路径中的所有反应都在同一反应器中连续进行，生产工艺简单，过程条件容易控制；②生物质来源广泛，价格低廉，将其通过化学手段转化为具有高附加值的乙酰丙酸酯，能够变废为宝；③直接采用生物质经醇解制取乙酰丙酸酯，反应路径短，无需对中间产物乙酰丙酸进行提纯，大大减少操作费用，从而提高技术效率；④反应完成后，根据体系中物质沸点的不同，产物乙酰丙酸酯容易从反应混合物中通过蒸馏分离获得，剩余未反应的醇可以回收循环使用。因此，利用木质纤维素醇解制备乙酰丙酸酯是目前的研究热点之一。

甲醇、乙醇和丙醇常被用作醇解木质纤维素制取乙酰丙酸酯的反应溶剂。在醇溶剂中加入适量的水可以促进木质纤维素转化为乙酰丙酸酯[184]，这可能是因为水的加入能够促进纤维素的水解反应和5-烷氧基甲基糠醛的水解开环反应。与木质纤维素制备乙酰丙酸相同，木质纤维素醇解制备乙酰丙酸酯同样需要酸性催化剂，包括液体酸（如硫酸、盐酸、磷酸、对甲苯磺酸等）、固体酸（如酸性树脂、分子筛、磺化分子筛、磺化碳材料、磺化金属氧化物等）和金属盐（硫酸铜、硫酸铝、氯化铁、硫酸铁等）。其中，硫酸和硫酸铝由于成本低、容易获取，且对生物质转化为乙酰丙酸酯具有较好的催化作用，因此得到了较广泛的应用和研究。特别地，Huang 等在甲醇中用 $Al_2(SO_4)_3$ 催化纤维素转化得到 70.6% 的乙酰丙酸甲酯[184]。

2.5 催化木质纤维素及其衍生碳水化合物转化制备乳酸（酯）的原理

2.5.1 乳酸简介

乳酸（lactic acid）是一种重要的碳水化合物衍生 α-羟基酸，可以用于制备可生物降解的高分子聚乳酸，或作为原料制备多种精细化学品[185,186]。如图 2-13 所示，乳酸可经脱水反应得到乙醛、丙烯酸、2,3-戊二酮，经加氢反应则得到 1,2-丙二醇，经氧化反应可以得到丙酮酸，或经酯化反应得到乳酸酯、丙交酯、聚乳酸等[2,187,188]。这些乳酸衍生物可以作为有机化工原料或者聚酯单体用于生产可再生精细化学品、有机溶剂和高聚物。聚乳酸是一种脂肪族聚酯和生物相容性热塑性材料，是目前最有发展前景和最受欢迎的材料，被认为是"绿色"环保材料，可用于药物控制释放、植入性复合材料、骨固定部件、包装和纸张涂层、农药和肥料缓释系统、堆肥袋等。

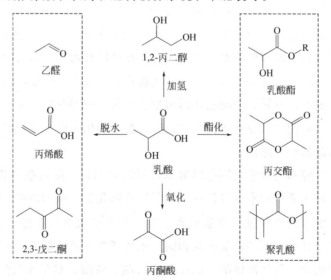

图 2-13 乳酸为原料制备的化学品[73,185]

目前工业上采用生物发酵的方法生产乳酸，该方法只能采用葡萄糖、淀粉作为原料，而不能采用廉价且在自然界中广泛存在的木质纤维素作为原料。近年的研究发现，通过化学催化的方法能够将纤维素甚至是木质纤维素原料高选择性地转化为乳酸（酯）。由于乳酸作为平台化学品的广泛用途，国内外研究人员对通过化学催化转化碳水化合物制备乳酸（酯）开展了大量的研究，开发了诸多高效的催化反应体系。本章主要对化学催化碳水化合物制备乳酸（酯）的研究进行论述，不涉及生物发酵法制备乳酸的研究。

2.5.2 碳水化合物制备乳酸（酯）的反应路径

如图 2-14 所示，葡萄糖水热降解生成乳酸的反应路径及所涉及的基本反应包括[189]：葡萄糖在水热条件下（特别是碱性条件）发生 1,2-氢转移和逆羟醛缩合反应生成 1,3-二羟基丙酮和甘油醛，其中 1,3-二羟基丙酮继续发生 1,2-氢转移转化为甘油醛，而甘油醛则经 β-消除生成丙酮醛，并进而发生坎尼扎罗反应生成乳酸。这条催化水热转化己糖生成乳酸的路径已经被目前大多数的研究者接受[78,190-197]。不过，Tolborg 等[198] 认为葡萄糖在水热转化为乳酸的过程中，葡萄糖也可能先发生 β-消除反应而生成 3-脱氧葡萄糖醛酮，3-脱氧

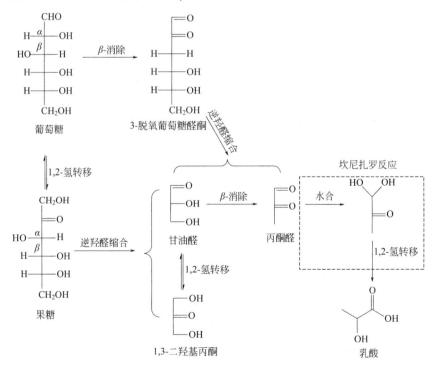

图 2-14　葡萄糖生成乳酸的反应路径

葡萄糖醛酮则发生逆羟醛缩合反应而生成甘油醛和丙酮醛,而这两种 C_3 化合物再继续转化为乳酸。

当在水热条件下转化己糖制备乳酸(酯)时,易发生下述副反应:①葡萄糖发生逆羟醛缩合反应生成赤藓糖和乙醇醛[199];②果糖发生脱水反应生成 HMF,并进而降解为乙酰丙酸[77];③葡萄糖、果糖及其中间产物发生聚合反应,生成固体残渣,即水热焦炭[200]。这些副反应的发生会降低己糖在生成乳酸(酯)过程中的目标产物的选择性。

2.5.3 反应原料及反应体系

可通过化学催化方法生产乳酸(酯)的碳水化合物原料包括三碳糖(1,3-二羟基丙酮和甘油醛)、己糖(葡萄糖和果糖)、二聚己糖(主要为蔗糖)、纤维素和含有纤维素的生物质。采用三碳糖为原料制备乳酸(酯)的反应由于涉及的反应步骤少,所以反应条件温和(反应温度通常在100℃左右),产物选择性高(往往达到60%以上),但是这些三碳糖在自然界中含量少,价格较为昂贵;采用可溶性己糖(葡萄糖、果糖和蔗糖)制备乳酸(酯)涉及己糖的逆羟醛缩合反应,且副产物(HMF、乙酰丙酸酯等)相对较多,所以产物选择性相对较低,而反应温度通常为180℃左右;由于纤维素转化为乳酸(酯)的过程中涉及纤维素水解为葡萄糖这一需要在较苛刻条件下进行的反应,因此利用纤维素及生物质原料制备乳酸(酯)的过程反应温度通常达到190℃以上,而乳酸(酯)的选择性也相对较低。

2.5.4 碳水化合物转化为乳酸(酯)的催化剂

由于碳水化合物转化为乳酸(酯)涉及多步反应,这就需要催化剂能够对每一步反应都有较好的催化作用。因此,催化剂的开发是利用碳水化合物制备乳酸(酯)的研究热点。目前已经发现了大量能够催化碳水化合物转化为乳酸(酯)的催化剂,根据催化剂的性质,可以把催化剂分为:无机碱及碱性无机盐、可溶性金属盐和固体酸。

2.5.4.1 无机碱及碱性无机盐

因为逆羟醛缩合反应和坎尼扎罗反应是己糖转化为乳酸(酯)的关键步骤,而碱性催化剂对这两个步骤都有较好的催化作用[190],所以部分研究采用强碱如 NaOH、$Ca(OH)_2$、$Ba(OH)_2$、$Sr(OH)_2$[201-210],或者一些碱性无机盐如镁铝水滑石[208]、Na_2SiO_3[211],来催化葡萄糖或纤维素的转化制备乳酸,通常能够得到30%～50%的乳酸。

碳水化合物水热转化过程中会生成大量水热焦炭，从而降低该过程中碳原子的利用率。部分研究发现，轻度氧化是抑制水热焦炭生成的有效方法[172,212]。Onda等[212]在80℃的NaOH水溶液中，采用流动的空气作为氧化剂，用Pt/Al_2O_3或Pt/MgO催化葡萄糖转化可得57%的乳酸并联产20%～40%的葡萄糖酸。该体系中，NaOH的作用是催化葡萄糖经异构化、逆羟醛缩合、β-消除、坎尼扎罗反应而生成乳酸，而Pt/Al_2O_3和Pt/MgO的作用是催化葡萄糖发生氧化反应而生成葡萄糖酸，从而抑制固体水热焦炭的产生。这一研究与Lin等[172]在氧化环境中催化纤维素轻度氧化转化生成乙酰丙酸的思路有异曲同工之处。

需要注意，这些用无机碱作催化剂催化碳水化合物制备乳酸的过程中，使用的碱催化剂会与生成的乳酸发生中和反应而生成乳酸盐，导致无机碱催化剂无法回收，而乳酸再生过程又会消耗大量无机酸并副产无机盐。此外，这些碱性催化剂在高温下往往对设备产生较强的腐蚀。因此，虽然这些碱性催化剂价格便宜、易于获取，但是实现大规模工业应用仍较为困难。

2.5.4.2 可溶性金属盐

2005年，锡盐和锌盐分别被用于催化碳水化合物在水热条件下转化为乳酸（酯）。Hayashi等[213]在90℃的醇溶剂（甲醇、乙醇、丁醇）中用$SnCl_4$、$SnCl_2$催化三碳糖（1,3-二羟基丙酮和甘油醛）的转化，最高可得到89%的乳酸（酯）；Bicker等则在亚临界水热环境中用$ZnSO_4$催化二羟基丙酮、葡萄糖、蔗糖转化，分别得到86%、42%和48%的乳酸[214]。随后，人们发现大量的可溶性金属盐可以催化碳水化合物转化为乳酸（酯），包括Pb^{2+}盐[215]、Er^{3+}盐[216,217]、Zn^{2+}盐、VO^{2+}盐[218]、Al^{3+}-Sn^{2+}二元盐[219]等。这些金属盐可以分为两性金属盐（Zn^{2+}、Co^{2+}、Al^{3+}、Cr^{3+}、Pb^{2+}、Ni^{2+}、In^{3+}、VO^{2+}、Sn^{4+}）和镧系金属盐（Er^{3+}、La^{3+}）。表2-3列出了可溶性金属盐催化碳水化合物制备乳酸（酯）的催化剂及其反应体系。

表2-3 可溶性金属盐催化碳水化合物制备乳酸（酯）的催化剂及其反应体系

原料	溶剂	催化剂	反应条件	乳酸(酯)收率/%	参考文献
三糖(1,3-二羟基丙酮、甘油醛)	甲醇、乙醇、正丁醇	$SnCl_2$	90℃,1～6h	78～89	[213]

续表

原料	溶剂	催化剂	反应条件	乳酸(酯)收率/%	参考文献
三糖(1,3-二羟基丙酮、甘油醛)	甲醇、乙醇、正丁醇	$SnCl_4 \cdot 5H_2O$	90℃,1～3h	81～91	[213]
己糖(葡萄糖、果糖、蔗糖)	甲醇	$SnCl_4 \cdot 5H_2O$	160℃,2.5h	28	[220]
己糖(葡萄糖、果糖、蔗糖)	甲醇	$NaOH+SnCl_4 \cdot 5H_2O$	160℃,2.5h	47～57	[220]
纤维素	甲醇	$SnCl_2$	190℃,4h	21.6	[221]
纤维素	甲醇	$SnCl_2+ZnCl_2$	210℃,4h	32.1	[221]
蔗糖	甲醇	$SnCl_2$	130℃,2h	23.5	[221]
蔗糖	甲醇	$SnCl_2+KCl$	130℃,2h	43.3	[221]
己糖(果糖、葡萄糖)	甲醇	$InCl_3+SnCl_2+NaBF_4$	160℃,10h	53～72	[222]
葡萄糖	甲醇	2-溴吡啶+$SnCl_2 \cdot 2H_2O$	220℃,6h	44.6	[225]
果糖、葡萄糖	水-ChCl	$SnCl_2$	190℃,1.5h	47～59	[226]
果糖、葡萄糖	水	Al^{3+}-Sn^{2+}	190℃,2h	81～90	[219]
纤维素	水	Al^{3+}-Sn^{2+}	190℃,2h	65	[219]
己糖(葡萄糖、果糖、蔗糖)	水	$ZnSO_4$	300℃,5～20s	42～48	[214]
己糖(葡萄糖、果糖、蔗糖)	乙醇	$ZnCl_2$	200℃,3h	47～52	[227]
葡萄糖	甲醇	$InCl_3$	170℃,6h	46	[223]
纤维素	水	$Pb(NO_3)_2$	190℃,4h	70	[215]
果糖、葡萄糖	水	$VOSO_4$	160℃,1.5h	56～58	[218]
纤维素	水	$VOSO_4$	180℃,4h	24	[218]
纤维素	水	$Er(OTf)_3$	240℃,0.5h	89.6	[217]
纤维素	水	$ErCl_3$	240℃,0.5h	91.1	[216]
左旋葡聚糖、葡萄糖、纤维素	甲醇	$La(OTf)_3$	250℃,1h	73～75	[224]

在 $SnCl_2$ 催化己糖降解生成乳酸(酯)的过程中，由于 $SnCl_2$ 水解产生的 HCl 能够催化己糖发生脱水反应而生成 HMF 和乙酰丙酸，所以仅用 $SnCl_2$ 催化己糖降解生成的乳酸(酯)收率通常仅 20%～30%[191,220,221]。部分研究尝试添加其他助催化剂(例如 NaOH、NaCl、KCl、$MgCl_2$、$InCl_3$、$AlCl_3$、2-

溴吡啶、氯化胆碱等）以提高 $SnCl_2$ 对己糖转化为乳酸（酯）的催化活性[220-222]。Deng 等[219] 报道了一种由 Al^{3+}-Sn^{2+} 组成的二元金属盐催化体系，能够催化果糖、葡萄糖和纤维素转化分别生成 90%、81% 和 65% 的乳酸（酯）。他们认为，该反应体系中的 Al^{3+} 能够催化反应过程中涉及的 1,2-氢转移反应，而 Sn^{2+} 主要是对己糖的逆羟醛反应起催化作用。

除了研究较多的锡盐，部分两性金属盐也被报道对碳水化合物特别是己糖转化为乳酸（酯）具有催化作用。Zn、In、Pb 和 V 是几种典型的两性金属，而它们的盐类都被报道对碳水化合物转化为乳酸（酯）具有较好的催化作用。Bicker 等[214] 发现 $ZnSO_4$ 对催化碳水化合物转化为乳酸具有催化作用。Wang 等报道了 $Pb(NO_3)_2$ 在 190℃ 的高温水中催化转化纤维素可以得到近 70% 的乳酸[215]。最近，Lu 等[223] 报道了 $InCl_3$ 在 170℃ 的甲醇溶剂中催化葡萄糖和果糖转化，分别可得到 46% 和 57% 的乳酸甲酯。Tang 等报道了 $VOSO_4$ 催化果糖和葡萄糖转化得到 56%～58% 的乳酸，当采用纤维素为原料时，也能够得到 24% 的乳酸[218]。因为大量两性金属盐都对催化碳水化合物生成乳酸（酯）具有较好的催化作用，笔者推测，极有可能这些两性金属盐在催化过程中具有相似的催化作用机理。

部分镧系金属（Er、La）的盐类对催化己糖制备乳酸（酯）也表现出较好的催化作用。Wang 等发现 Er 的盐类如三氟甲基磺酸铒 $[Er(OTf)_3]$ 及氯化铒（$ErCl_3$），都对葡萄糖转化为乳酸具有较好的催化效果[216,217]。他们采用 $Er(OTf)_3$ 作为催化剂在 240℃ 的热水中转化纤维素 0.5h，能够得到 89% 的乳酸[217]，而采用 $ErCl_3$ 作为催化剂在 240℃ 的热水中转化纤维素甚至能够得到 91.1% 的乳酸[216]。Liu 等[224] 则报道了采用 $La(OTf)_3$ 于 250℃ 的甲醇中催化左旋葡聚糖、葡萄糖和纤维素转化，反应 1h 可得 73%～75% 的乳酸甲酯。

大量的研究都表明，Sn^{4+} 盐、Zn^{2+} 盐、Pb^{2+} 盐、In^{3+} 盐、VO^{2+} 盐、Er^{3+} 盐和 La^{3+} 盐都对催化葡萄糖甚至纤维素这些在自然界中来源广泛、价格低廉的碳水化合物制备乳酸（酯）这一过程表现出优异的催化性能。但是这些金属盐都存在难以与产物分离等不足，部分金属盐还存在价格昂贵（稀土金属盐）或者毒性大（Pb^{2+} 盐）等缺点，这些不足之处大大限制了金属盐在催化碳水化合物制备乳酸（酯）中的应用。相比之下，Zn^{2+} 盐和 Sn^{4+} 盐的价格低廉且毒性较低，更有可能在催化葡萄糖、纤维素甚至生物质等来源广泛且价格低廉的碳水化合物生成乳酸（酯）的领域实现工业应用。

2.5.4.3 固体酸催化剂

2007年，比利时的Sels课题组首次尝试用Y型分子筛固体酸催化剂催化三碳糖制备乳酸（酯）[228]。2009年，Taarning等[229]首次将Sn-β分子筛用于催化碳水化合物转化。他们发现Sn-β分子筛在80℃的甲醇中能够催化三碳糖转化得到99%的乳酸甲酯；当以水为溶剂时，乳酸的收率同样能够达到90%。随后，Holm等[191]以Sn-β分子筛为催化剂，分别以水和甲醇为反应介质转化蔗糖，得到约30%的乳酸和68%的乳酸甲酯。这是首次用固体酸催化己糖转化得到较高收率的乳酸酯。这一贡献大大推进了利用生物质原料制备乳酸（酯）领域的研究进展。随后，国内外在利用固体酸催化碳水化合物水热降解制备乳酸（酯）方面做了大量研究工作，开发出了以Sn-Si分子筛、两性金属氧化物为代表的固体催化剂。表2-4列举了一些近年来采用固体酸催化碳水化合物制备乳酸（酯）的研究成果。

表2-4 固体酸催化碳水化合物制备乳酸（酯）

原料	溶剂	催化剂	反应条件	乳酸(酯)收率/%	参考文献
二羟基丙酮、甘油醛	甲醇	Sn-β	80℃,24h	99	[229]
蔗糖	甲醇	Sn-β	160℃,2h	64	[191]
葡萄糖	甲醇	Sn-MCM-41	160℃,20h	43	[234]
菊芋	甲醇	Sn-SBA-15	160℃,20h	57	[238]
蔗糖	甲醇	Sn-Si-CSM	155℃,20h	45	[240]
葡萄糖、蔗糖	甲醇	M-Sn-β(M=Li,Na,K,Rb,Cs,Zn,Pb)	170～190℃,2～16h	52～65	[241-243]
蔗糖	甲醇	Sn-β+K_2CO_3	170℃,16h	75	[241]
葡萄糖	甲醇	Sn-β+KX(X=Cl,Br,I,NO_3)	120℃,19h	45～55	[245]
果糖	乙醇	Sn-MFI+(MoO_3,MoO_2 或 MoS_2)	100℃,20h	69.2	[246]
蔗糖	水	Sn-β-NH_2	190℃,4h	58	[244]
木糖 木聚糖	水	ZrO_2	200℃,40～90min	30～42	[254]
纤维素	水	ZrO_2	200℃,6h	21.2	[262]
纤维素	水+乙醇	Zr-SBA-15	260℃,6h	33	[255]
纤维素	水	ZrO_2-Al_2O_3	200℃,6h	25.3	[263]
葡萄糖	水	ZnO/SiO_2	180℃,1h	39.2	[247]

续表

原料	溶剂	催化剂	反应条件	乳酸(酯)收率/%	参考文献
纤维素	甲醇	Ga-ZnO/HNZY	260℃,2h	58	[257]
葡萄糖	甲醇	γ-Al$_2$O$_3$	160℃,6h	34	[264]
果糖	甲醇	NiO	200℃,3h	47	[265]
葡萄糖	水	Nb$_2$O$_5$	250℃,4h	38	[248]
果糖	水	Pb(OH)$_2$/rGO	190℃,2h	58.7	[253]
纤维素	水	Al-W	190℃,24h	30	[30]
纤维素	水	Al$_2$(WO$_4$)$_3$	220℃,3h	40	[249]
纤维素	水	Er/K10	240℃,30min	67.6	[259]
纤维素	水	Er$_2$O$_3$/Al$_2$O$_3$	240℃,3h	45.8	[260]
纤维素	水	Er/β	240℃,30min	58	[258]
果糖	水	La-HPMo-3	170℃,4h	65	[261]

由于Sn-β分子筛在催化碳水化合物（特别是己糖）转化为乳酸（酯）过程中表现出优异的催化性能，后续大量工作对此开展了研究。部分研究尝试对Sn-β分子筛的结构进行改良，如制备包含微孔/介孔的Sn-β分子筛[230,231]、纳米Sn-β分子筛[232]。其他孔结构的锡硅酸盐的分子筛如Sn-Mont[233]、Sn-MCM-41[234,235]、Sn-USY[236]、Sn-WMM[237]、Sn-SBA-15[238]等，甚至是Sn-Si混合氧化物[239]，也都被用于催化己糖或纤维素转化为乳酸（酯），获得40%~70%的乳酸（酯）。这些研究表明，孔道结构虽然对Sn-Si催化剂的催化性能有一定的影响，但是这些孔道结构不是Sn-Si催化剂活性的主要因素。

部分研究将新的活性组分引入Sn-β分子筛，得到新的固体催化剂如Sn-Si-CSM、Li-Sn-β、Na-Sn-β、Rb-Sn-β、Cs-Sn-β、Zn-Sn-β、Pb-Sn-β和Sn-β-NH$_2$等[240-244]，并发现新组分的引入都能提高催化剂对碳水化合物转化为乳酸（酯）的催化活性。也有的研究尝试将掺锡分子筛与其他助催化剂（例如K$_2$CO$_3$、KCl、KBr、KI、KNO$_3$、MoO$_3$）进行组合[241,245,246]，以实现协同催化碳水化合物转化为乳酸（酯），特别地添加碱金属盐类对提高Sn-β的催化活性具有显著的效果。这可能是由于碱金属离子对逆羟醛缩合反应具有较好的催化作用。

除了掺杂锡的硅酸盐分子筛，研究还发现部分两性金属氧化物（ZrO$_2$、ZnO、NiO、Al$_2$O$_3$、Fe$_2$O$_3$、SnO$_2$、Nb$_2$O$_5$、PbO、Cr$_2$O$_3$等）也对碳水化合物转化为乳酸（酯）具有一定的催化作用，并研制出一系列以两性金属氧化物为活性中心的固体酸催化剂，如 ZnO/SiO$_2$[247]、Fe$_2$O$_3$-SnO$_2$[248]、SnO$_2$-SiO$_2$[235,239]、Al$_2$O$_3$-WO$_3$[30,249]、Nb$_2$O$_5$-SiO$_2$[250]、SnO$_2$-P$_2$O$_5$[251]、Cr$_2$O$_3$-TiO$_2$/SiO$_2$[252]、Pb(OH)$_2$/rGO[253] 等，并发现这些催化剂都对催化碳水化合物转化为乳酸（酯）具有一定的催化作用。Lin 等[254] 将 Zr^{4+} 掺杂到 SBA-15 中，得到的 Zr-SBA-15 分子筛能够在 260℃ 的水-乙醇溶剂体系中直接转化纤维素得到 33% 的乳酸乙酯[255]。此外，该催化剂能够在水-甲醇溶剂体系中催化果糖、葡萄糖及纤维素转化，分别得到 44.1%、37% 和 28% 的乳酸甲酯[256]。他们认为，这些含锆固体酸中的 O—Zr—O 结构作为路易斯酸中心，可能与糖的羰基发生作用，从而触发己糖的逆羟醛反应。Verma 等[257] 合成了纳米 Ga-ZnO/HY 催化剂，能在 280℃ 的超临界甲醇溶液中催化微晶纤维素转化得到 57.8% 的乳酸甲酯，并副产 12.8% 的 2-甲氧基丙酸甲酯。他们认为，该催化剂中 ZnO 上的 Zn^{2+} 作为路易斯酸性位点是催化葡萄糖异构化反应及果糖逆羟醛反应的活性位点，而掺杂的 Ga^{2+} 与 ZnO 相互作用，产生新的路易斯酸位点对果糖逆羟醛反应具有重要的催化作用。

因为镧系金属的可溶性盐对碳水化合物转化为乳酸具有催化作用，部分研究开发了以镧系金属（Er、Yb）为活性中心、对碳水化合物转化为乳酸（酯）具有较好催化作用的固体酸。因前期研究发现 Er^{3+} 的可溶性盐类对催化碳水化合物转化为乳酸具有良好的催化作用[216,217]，Dong 等通过多种方法制备了含有铒元素的固体酸，如将 Er^{3+} 交换到蒙脱土 K10、β 分子筛[258] 及 ZSM-5 分子筛上[259]，或者将 Er$_2$O$_3$ 负载在 Al$_2$O$_3$ 上[260]，所得到的固体催化剂均能够催化转化纤维素分别得到收率为 50% 以上的乳酸[258-260]。Zhao 等[261] 采用离子交换法制备了镧改性的 La-HPMo-3，发现该催化剂能够在 170℃ 下转化果糖得到 65% 的乳酸。

2.5.5 碳水化合物水热转化生成其他 α-羟基酸

除了乳酸，碳水化合物在水热转化过程中还可以生成碳链长度为 4～6 的 α-羰基醛，并进而转化为相应的 α-羟基酸（酯），如 2-羟基-3-丁烯酸（酯）、2,4-二羟基丁酸、2,5-二羟基-3-戊烯酸、2,4,5-三羟基戊酸、2,4,5,6-四羟基己酸等。表 2-5 列出了几种生物质衍生单糖（葡萄糖、木糖、赤藓糖和甘油醛）在水热降解过程中生成的 α-羰基醛和 α-羟基酸。可以看到，催化转化碳水化合物水热转化生成 C$_3$ 的 α-羟基酸只有乳酸，而 C$_4$～C$_6$ 的 α-羟基酸都包括两种，即 4-羟基结构的 α-羟基酸和 3-烯基结构的不饱和 α-羟基酸。与乳酸相同,这些 α-羟基酸可以用于制备可再生的聚酯,或者用于生成诸多高附加值的化学品[266,267]。

表 2-5 醛糖及其所生成的 α-羰基醛和 α-羟基酸

醛糖	α-羰基醛	α-羟基酸	β,γ-不饱和 α-羰基醛	β,γ-不饱和 α-羟基酸
葡萄糖	3-脱氧葡萄糖酮醛	2,4,5,6-四羟基己酸	5,6-二羟基-2-氧代-3-己烯醛	2,5,6-三羟基 3-己烯酸
木糖	4,5-二羟基-2-氧代戊醛	2,4,5-三羟基戊酸	5-羟基-2-氧代-3-戊烯醛	2,5-二羟基-3-戊烯酸
赤藓糖	4-羟基-2-羰基丁醛	2,4-二羟基丁酸	2-羰基-3-丁烯醛	2-羰基-3-丁烯酸
甘油醛	丙酮醛	乳酸	—	—

关于碳水化合物在水热条件下转化 α-羟基酸(酯)的路径,目前已经达成了共识(图 2-15)[266,268,269]:醛糖(葡萄糖、木糖、赤藓糖、甘油醛)通过 β-消除反应和酮式-烯醇互变反应而转化为相应的 α-羰基醛,而 α-羰基醛则经分子内坎尼扎罗反应转化为 α-羟基酸。另外,当 α-羰基醛的 γ-C 原子上有羟基时,α-羰基醛可继续发生 β-消除反应生成 β,γ-不饱和-α-羰基醛,而 β,γ-不饱和-α-羰基醛则继续发生分子内坎尼扎罗反应生成 β,γ-不饱和-α-羟基酸。当上述反应在甲醇、乙醇、丙醇、丁醇等醇类溶剂中发生时,生成的 α-羟基酸可与溶剂反应生成相应的酯类。

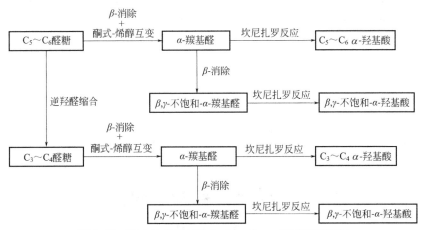

图 2-15 碳水化合物转化为 $C_3 \sim C_6$ α-羟基酸的路径

其中,葡萄糖经 β-消除和酮式-烯醇互变生成 3-脱氧葡萄糖酮醛,3-脱氧葡萄糖酮醛则经坎尼扎罗反应生成 2,4,5,6-四羟基己酸;此外,3-脱氧葡萄糖酮醛还可以继续发生 β-消除反应生成 5,6-二羟基-2-氧代-3-己烯醛,而 5,6-二羟基-2-氧代-3-己烯醛则发生坎尼扎罗反应而生成 2,5,6-三羟基-3-己烯酸[198,270]。木糖经 β-消除和酮式-烯醇互变生成 4,5-二羟基-2-氧代戊醛，而 4,5-二羟基-2-氧代戊醛则发生坎尼扎罗反应而生成 2,4,5-三羟基戊酸，或者继续发生 β-消除反应而生成 5-羟基-2-氧代-3-戊烯醛，并进而发生坎尼扎罗反应而生成 2,5-二羟基-3-戊烯酸[270,271]。赤藓糖经 β-消除和酮式-烯醇互变而生成 4-羟基-2-羰基丁醛，4-羟基-2-羰基丁醛可以发生坎尼扎罗反应而生成 2,4-二羟基丁酸，或者继续发生 β-消除而生成 2-羰基-3-丁烯酸，而 2-羰基-3-丁烯醛可以发生坎尼扎罗反应而生成 2-羟基-3-丁烯酸[191,268,270,272,273]。甘油醛经 β-消除和酮式-烯醇互变生成丙酮醛，丙酮醛经坎尼扎罗反应生成乳酸。由于丙酮醛的缺少 γ-C，所以丙酮醛不能发生 β-消除而生成 β,γ-不饱和-α-羰基醛和 β,γ-不饱和-α-羟基酸。

2.6 纤维素制备己糖醇（山梨醇和甘露醇）

己糖（葡萄糖和果糖）分子上的羰基易被还原为羟基，所以己糖可以在加氢条件下转化为己糖醇，如山梨醇和甘露醇。从纤维素到己糖醇的过程涉及纤维素水解和葡萄糖加氢两个步骤，其中，纤维素水解过程需要酸催化剂，而葡萄糖加氢步骤需要金属催化剂，因此，将木质纤维素转化为己糖醇通常需要酸催化剂和金属催化剂的共同作用（图 2-16）。近十年来国内外在纤维素催化氢解制备山梨醇方面开展了大量研究，具体可以见相关综述[274,275]。表 2-6 列出了部分利用纤维素制备己糖醇的催化剂及反应参数。

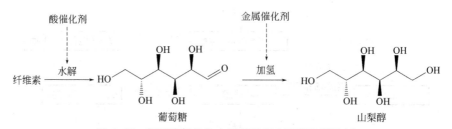

图 2-16 纤维素转化为山梨醇的反应路径及催化剂

表 2-6 利用纤维素制备己糖醇的反应体系

实验原料	催化剂	反应条件	己糖醇收率/%	参考文献
微晶纤维素	Ru/C	245℃,6MPa H_2,0.5h	39.3	[276]
浓磷酸处理的纤维素	Ru/CNT	185℃,5MPa H_2,24h	69	[277]
球磨纤维素	Ru/AC	205℃,5MPa H_2,5h	68.8	[278]

续表

实验原料	催化剂	反应条件	己糖醇收率/%	参考文献
球磨纤维素	Ru-Ni/AC	205℃,5MPa H_2,5h	69.4	[280]
球磨纤维素	Ni/CNF	230℃,6MPa H_2,4h	56.5	[281]
微晶纤维素	Ni_2P/AC	225℃,6MPa H_2,1.5h	53.1	[282]
球磨微晶纤维素	Ni/CNF	190℃,6MPa H_2,24h	76	[283]
微晶纤维素	H_2SO_4+Ru/C	160℃,5MPa H_2,1h	33.2	[284]
球磨纤维素	HCl+Ru/H-USY	190℃,5MPa H_2,24h	93	[287]
球磨纤维素	H_2SO_4+Ru/C	150℃,5MPa H_2,1h	94.3	[286]
微晶纤维素	$H_4SiW_{12}O_{40}$+Ru/C	190℃,5MPa H_2,1.5h	38	[288]
球磨纤维素	$H_4SiW_{12}O_{40}$+Ru/C	190℃,9.5MPa H_2,1h	85	[288]
球磨纤维素	$Cs_{3.5}SiW$+Ru/C	190℃,5MPa H_2,48h	90	[289]
微晶纤维素	Pt/γ-Al_2O_3	190℃,5MPa H_2,24h	31	[290]
球磨纤维素	Ru/$Cs_3PW_{12}O_{40}$	160℃,2MPa H_2,24h	45	[293]
球磨纤维素	Ru/SiO_2-SO_3H	150℃,4MPa H_2,10h	61.2	[292]
球磨纤维素	Ru/$NbOPO_4$	170℃,4MPa H_2,24h	69	[294]
微晶纤维素	Ni/ZSM-5	240℃,4MPa H_2,4h	60	[295]

纤维素的水解反应通常需要在较高的温度下进行,而除了碳材料之外的大多数固体催化剂的水热稳定性都较差,因此,碳材料是高温氢解纤维素优先考虑的催化剂载体。另一方面,在高温热水中,水分子可以电离出一定量的质子从而催化纤维素的水解反应,因此直接将金属负载于中性碳材料上得到的固体催化剂(例如 Ru/C、Ru/CNT、NiP/C、Ni/CNF)能催化纤维素转化为己糖醇[276-283]。Luo 等在高温热水中用 Ru/C 催化纤维素氢解能得到 39% 的己六醇[276]。为了克服金属-碳催化剂对纤维素水解的催化作用较弱的问题,部分研究在纤维素转化之前对纤维素进行预处理(浓酸处理或球磨处理)以提高纤维素与催化剂的接触面积。Deng 等将 Ru 负载于碳纳米管上制备出 Ru/CNT 并将其用于催化磷酸处理后的纤维素进行氢解可得到 69% 的己糖醇[277]。Ribeiro 等将 Ru/AC 和 Ru/CNT 用于催化经球磨处理的纤维素的水热氢解,分别能够得到 80% 和 50% 的山梨醇[278,279]。上述研究都采用了贵金属作为加氢催化剂,为了节省贵金属,Ribeiro 等将 Ru-Ni/AC 和 Ru-Ni/CNT 催化剂用于水热氢解纤维素,能够得到 50%~60% 的山梨醇。他们认为,Ru 和 Ni 之间具有协同催化作用从而提高山梨醇的选择性[280]。Van De Vyver 等将 Ni 负载于碳纳米纤维上,制备的 Ni/CNF 能够催化球磨处理后的纤维素转化得到

50%的山梨醇[281]。

催化剂中的酸性一方面可以提高对纤维素水解的催化作用，另一方面可以抑制葡萄糖的逆羟醛缩合反应。因此，有的研究将金属负载于活性炭（Pt/C、Ru/C、Pd/C）上，然后将该催化剂与酸性催化剂（H_2SO_4、HCl、$H_4SiW_{12}O_{40}$、$Cs_{3.5}SiW$ 等）组成二元催化剂，实现纤维素转化为己糖醇[284-289]。Palkovits 等采用 H_2SO_4 和 Ru/C 协调催化纤维素水热氢解，能够得到 60%的己六醇[284]。Geboers 等将 HCl 和 Ru/H-USY 组合催化球磨纤维素水热氢解，得到 93%的己六醇[287]。随后他们使用 $H_4SiW_{12}O_{40}$ 和 Ru/C 催化剂体系去催化球磨纤维素水热氢解，能够得到 92%的己六醇[288]。可以看到，该类反应体系对转化纤维素制备己糖醇具有较好的催化作用。

有研究将金属负载于酸性载体上，得到金属-酸双功能催化剂（例如 Pt/Al_2O_3[290]、Ru/AC-SO_3H[291]、Ru/SiO_2-SO_3H[292]、Ru/$Cs_3PW_{12}O_{40}$[293]、Ru/SiO_2-Al_2O_3、Ru/$NbOPO_4$[294]、Ni/ZSM-5[295]），并将该类催化剂用于催化纤维素氢解制备己糖醇。Zhu 等使用 Ru/SiO_2-SO_3H 催化纤维素水热氢解生成 61.2%的山梨醇[292]。Liu 等将 Ru/$Cs_3PW_{12}O_{40}$ 用于转化纤维素得到 43%的山梨醇[293]。Negoi 等将贵金属 Ir、Pd、Rh 和 Ru 负载于 BEA 沸石上并将其用于纤维素的水热氢解，发现 Ru-BEA 能够催化纤维素氢解得到 72.5%的山梨醇[296]。Xi 等采用 Ru/$NbOPO_4$ 催化纤维素水热氢解得到 59%～69%的山梨醇[294]。这类催化剂的问题在于催化剂载体的水热稳定性较差，使得催化剂在多次利用后活性显著下降。提高该类催化剂载体的水热稳定性，或者探寻失活催化剂的再生方法，对于提高这类催化剂的工业应用具有一定的意义。

2.7 纤维素制备乙二醇和 1,2-丙二醇

乙二醇和 1,2-丙二醇是两种市场需求量很大的高价值化学品，广泛用于合成聚酯纤维、不饱和树脂、防冻剂和其他精细化学品。随着全球对服装、包装和不饱和树脂材料需求的增长，市场容量预计在未来 10～20 年内将以每年 5%的速度进一步增长。目前，乙二醇和 1,2-丙二醇主要由石油行业衍生的乙烯和丙烯通过环氧化反应生产，或者从煤化工行业生产。2008 年，大连化物所张涛院士团队在研究纤维素氢解时首次发现 Ni-W_2C/AC 可以高效地催化纤维素转化，并得到 61%的乙二醇和约 5%的 1,2-丙二醇[297]。在这一突破性地研究被报道之后，世界各地的许多小组进行了广泛的研究，以探索新的催化

剂，深入了解反应机理，调整反应选择性，模拟和识别反应动力学，并开发出了一系列能够高选择性地将木质纤维素转化为乙二醇和1,2-丙二醇的催化剂和反应体系[199,298,299]。

2.7.1 纤维素制备乙二醇和1,2-丙二醇的路径

图2-17展示了纤维素转化为乙二醇和1,2-丙二醇的路径。催化纤维素制备乙二醇的过程所涉及的反应步骤包括：①纤维素水解为葡萄糖；②葡萄糖发生逆羟醛缩合反应转化为赤藓糖和乙醇醛，赤藓糖再发生逆羟醛缩合反应转化为两个乙醇醛分子；③乙醇醛再经后续的加氢反应转化为乙二醇[199]。而纤维素转化为1,2-丙二醇的过程所涉及的反应步骤包括：纤维素水解为葡萄糖、葡萄糖异构化为果糖、果糖逆羟醛缩合反应生成1,3-二羟基丙酮和甘油醛、甘油醛脱水生成丙酮醛、丙酮醛加氢则得到1,2-丙二醇。所以，当催化剂对葡萄糖逆羟醛缩合反应的催化作用较强时，葡萄糖水热氢解的产物主要是乙二醇，当催化剂对葡萄糖异构化为果糖的催化作用较好时，葡萄糖水热氢解能够得到更多的1,2-丙二醇。

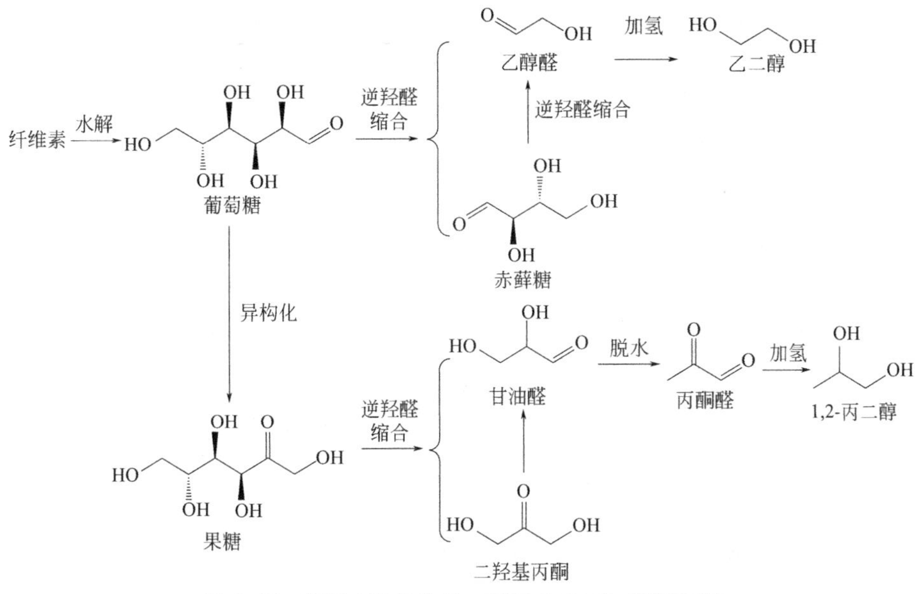

图 2-17　纤维素转化为乙二醇和1,2-丙二醇的路径

在纤维素水热氢解生成乙二醇和1,2-丙二醇的过程中，还会发生诸多副反应，比如，葡萄糖和果糖可直接加氢生成己糖醇，赤藓糖可直接加氢生成赤藓糖醇，二羟基丙酮和甘油醛可直接加氢生成丙三醇。

2.7.2 纤维素制备乙二醇和 1,2-丙二醇的催化体系

在水热催化纤维素氢解生成乙二醇和 1,2-丙二醇的过程中，一种有效的催化剂应该能够完成不同的功能，以实现对纤维素水解反应、糖的逆羟醛缩合反应、糖的异构化反应和醛/酮的加氢反应的催化作用。水解反应很容易在液体无机酸或固体酸的催化下进行，加氢反应可以在含有 Ni、Pd、Pt、Rh、Ir、Ru 或 Cu 金属的常规金属催化剂上进行，而糖的异构化可以在碱或路易斯酸存在的情况下发生[300]。纤维素催化转化为乙二醇的核心步骤是逆羟醛缩合反应，己糖通过该反应生成 C_2 和 C_3 中间体。由于上述几个反应发生在不同的活性中心，因此，有效的偶联反应可以在纤维素水热氢解过程中得到较高的乙二醇和 1,2-丙二醇收率。催化剂可以是含有多功能组分的单一催化剂，如 NiW_2C/AC、WC_x/MC、Ni-W/SBA-15、$Pt-SnO_x/Al_2O_3$，也可以是由两种或三种单功能催化剂组成的混合物，如 $Ru/AC+H_2WO_4$、$Raney\ Ni+H_2WO_4$、$Ru/C+AMT$ 和 $Ru/C+WO_3+AC$。表 2-7 列举了利用纤维素及生物质制备乙二醇和 1,2-丙二醇的反应体系。

表 2-7 利用纤维素及生物质制备乙二醇和 1,2-丙二醇的反应体系

原料	催化剂	反应条件	小分子醇收率/%		参考文献
			乙二醇	丙二醇	
纤维素	$Ni-W_2C/AC$	245℃,6MPa H_2,0.5h	61	7.6	[297,301]
纤维素	Ni-WP/AC	245℃,6MPa H_2,0.5h	46	6.4	[304]
纤维素	(Pd、Pt、Ir 或 Ru)-W/AC	245℃,6MPa H_2,0.5h	50.6~60.7	2.5~3.8	[305]
纤维素	Ni-W/SBA-15	245℃,6MPa H_2,0.5h	75.4	3.2	[305]
纤维素	$Ru/AC+H_2WO_4$	245℃,6MPa H_2,0.5h	58.5	3.5	[306]
纤维素	$Raney\ Ni+H_2WO_4$	245℃,6MPa H_2,0.5h	65.4	3.3	[307]
纤维素	$Ru/C+WO_3$	245℃,6MPa H_2,0.5h	48.9	7.4	[18]
纤维素	$Ru/C+WO_3/Al_2O_3+AC$	245℃,6MPa H_2,0.5h	16.6	30.7	[18]
纤维素	Pt/AlW	190℃,5MPa H_2,24h	—	~20%	[34]
纤维素	Ru/WO_3	240℃,4MPa H_2,2h	76.3	4.3	[308]
纤维素	$Al-WO_3-Ni-TUD-1$	230℃,6MPa H_2,1.5h	76	8.3	[35]
球磨纤维素	Ru-W/CNT	205℃,5MPa H_2,5h	38.5	7.8	[309]
纤维素	$Cu-Cr+Ca(OH)_2$	245℃,6MPa H_2,5h	31.6	42.6	[310]
葡萄糖	RuSn/AC	240℃,4MPa H_2,10min	26.9	25.0	[311]
桦木	$Ni-W_2C/AC$	235℃,6MPa H_2,4h	54.3	14.6	[302]
玉米秸秆	$Ni-W_2C$	245℃,6MPa H_2,2h	18.3	13.9	[303]

2008 年，张涛课题组的 Ji 等采用 Ni-W_2C/AC 为催化剂，在 518K 和 6MPa 的氢气压力下一步转化纤维素得到 61% 的乙二醇和约 5% 的 1,2-丙二醇，而生成的山梨醇则非常少[297,301]。上述催化体系对直接转化生物质原料生产乙二醇和 1,2-丙二醇同样具有较好的催化作用。他们将 Ni-W_2C/AC 用于催化转化未经处理的生物质如桦木、玉米秸秆等的水热氢解，最高可以得到 54% 的乙二醇（相对于纤维素和半纤维素），以及 46% 的酚类（相对于木质素）[302,303]。起初他们认为碳化钨是主要的加氢催化活性位，但是后续研究表明碳化钨的加氢作用其实比较弱[301,304]。随后他们发现，钨与其他金属共同组成的双金属加氢催化剂，如 Ru-W/AC、Pd-W/AC、Pt-W/AC、Ir-W/AC 和 Ni-W/SBA-15 等，都能催化纤维素转化为乙二醇并达到 50%～76% 的收率[305]。此外，采用金属加氢催化剂（Pt/C、Ru/C、Raney Ni）与各种含钨的化合物（磷钨酸、三氧化钨、硅钨酸、偏钨酸铵、钨酸）组成的二元催化体系也能有效催化纤维素降解得到 32%～65% 的乙二醇[18,306,307]。上述系列研究结果表明，在含有钨的催化剂转化纤维素生成乙二醇和 1,2-丙二醇的过程中，钨起到的主要作用是催化葡萄糖和果糖的逆羟醛缩合反应而不是催化加氢反应，这与 Chambon 等采用 AlW 和 Pt/AlW 作为催化剂转化纤维素分别得到乳酸和 1,2-丙二醇时钨所起到的作用是相同的[30,34]。

Liu 等研究了 Ru/C+WO_3 共同催化纤维素水热氢解，可以得到 48.9% 的乙二醇。他们发现，当只有 Ru/C 作为催化剂时，纤维素的转化率较低，且产物主要为己糖醇。而当采用 Ru/C+WO_3 共同催化纤维素的水热氢解时，纤维素的转化率显著提高，且乙二醇和 1,2-丙二醇的选择性也显著提高。这一结果表明，在 Ru/C+WO_3 共同催化纤维素水热氢解的过程中，WO_3 一方面起到催化纤维素水解的作用，另一方面还对葡萄糖的逆羟醛缩合反应起到催化作用。当他们在反应体系中添加活性炭后，乙二醇的收率下降而 1,2-丙二醇的收率显著上升，这是由于活性炭能够催化葡萄糖异构化为果糖[18]。

随着钨元素在纤维素水热氢解制备乙二醇中的作用被发现，大量后续研究采用含有钨的加氢催化剂去转化纤维素制备乙二醇。Li 等将纳米 Ru 负载在 WO_3 制备了 Ru/WO_3 并将其用于催化纤维素的水热氢解，能够得到 76.3% 的乙二醇[308]。Hamdy 等则将 Ni/AlW 用于催化纤维素水热解聚，在 230℃、4MPa H_2 条件下转化 1.5h，能得到 76% 的乙二醇[35]。他们指出，Al^{3+} 对纤维素的水解具有重要作用，钨原子则与葡萄糖的氧原子发生配位从而促进葡萄糖的逆羟醛缩合反应。Ribeiro 等通过多次浸渍法将 Ru 和 W 负载于碳纳米管上制备了 Ru-W/CNT，并发现它能够在 205℃、5MPa H_2 的条件下，于 3h 内催化纤维素完全降解并得到 38.5%的乙二醇和 7.8%的 1,2-丙二醇[309]。

因为纤维素转化为乙二醇涉及碳水化合物的逆羟醛缩合反应，所以部分研究将对逆羟醛缩合反应具有催化作用的催化剂与加氢催化剂结合来转化纤维素制备乙二醇和 1,2-丙二醇。Xiao 等采用 Cu-Cr 与 $Ca(OH)_2$ 共同催化纤维素的转化，得到 31.6%的乙二醇和 42.6%的 1,2-丙二醇。这一反应体系能够得到较高的 1,2-丙二醇可能是由于 Cr 可以催化葡萄糖异构化为果糖，$Ca(OH)_2$ 可以催化葡萄糖的逆羟醛缩合反应生成乙醇醛，而 Cu-Cr 催化乙醇醛的加氢反应[310]。张涛课题组的 Pang 等将 Ru 和 Sn 负载于活性炭上合成了 RuSn/AC 催化剂，用它催化葡萄糖水热氢解得到 25%的 1,2-丙二醇和 26.9%的乙二醇[311]。他们认为催化剂中的 SnO_2 能够催化葡萄糖异构化为果糖，而 Ru-Sn 合金则对碳水化合物的逆羟醛缩合反应起催化作用[311]。

2.8 碳水化合物在水热转化过程中发生的基本反应

前面几节介绍了碳水化合物生成 HMF、乙酰丙酸、乳酸、己糖醇、乙二醇和 1,2-丙二醇的基本原理。可以看到，葡萄糖水热降解生成 HMF 的过程所涉及的基本反应包括 β-消除反应、酮式-烯醇互变反应、缩醛环化反应和脱水反应等（图 2-6）；HMF 转化为乙酰丙酸的过程所涉及的基本反应包括脱水反应、呋喃环的水解开环反应、α-羰基醛的水解 C—C 键断裂反应、2-烯醛的水合重排反应（图 2-9）；葡萄糖生成乳酸的过程所涉及的基本反应包括 1,2-氢转移反应、逆羟醛缩合反应、α,β-羟基醛的 β-消除反应、α-羰基醛的坎尼扎罗反应（图 2-14）；葡萄糖生成乙二醇的过程涉及的反应包括逆羟醛缩合反应和加氢反应（图 2-17）。

对所有碳水化合物的水热降解路径进行分析，可以发现图 2-18 所示的一些基本规律：①葡萄糖和甘油醛都是 α-羟基醛，它们都可以经由 1,2-氢转移生成相应的酮糖，即果糖和 1,3-二羟基丙酮；②葡萄糖、木糖和甘油醛都是 α,β-二羟基醛，它们都通过经由 β-消除和酮式-烯醇互变而生成 α-羰基醛，分别是 3-脱氧葡萄糖醛酮、3-脱氧木糖醛酮和丙酮醛；③3-脱氧葡萄糖醛酮和丙酮醛都是 α-羰基醛，它们都可以经由坎尼扎罗反应生成相应的 α-羟基酸，分别是 2,4,5,6-四羟基己酸和乳酸；④3-脱氧葡萄糖醛酮和果糖的羰基相邻的 γ-C 原子上有羟基，这两种物质都可通过缩醛环化反应和脱水反应而生成具有呋喃环的物质；⑤葡萄糖、果糖的羰基相连的 α-C 和 β-C 原子上均有羟基，而这两种 α,β-二羟基醛都能够发生逆羟醛缩合反应。

所以，β-羟基醛的β-消除反应、α-羟基醛的1,2-氢转移、α-羰基醛的C—C键水解断裂反应、α-羰基醛的坎尼扎罗反应、α-羰基醛的羟醛缩合反应是碳水化合物降解过程中极易发生的基本反应，这些基本反应决定了碳水化合物在水热降解过程中的路径。

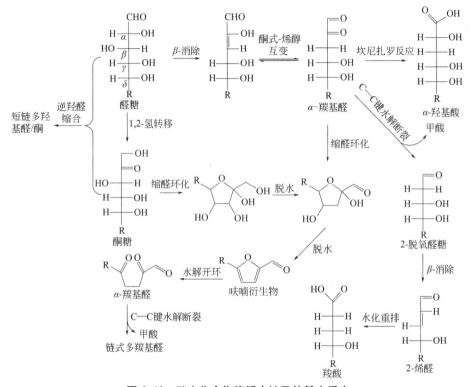

图2-18　碳水化合物降解中涉及的基本反应

对这些基元反应进行归纳，并研究催化剂对各种基元反应的催化作用机理，对于开发新型高效催化剂实现碳水化合物的定向转化具有重要意义。比如，因为葡萄糖转化为乳酸涉及逆羟醛缩合反应和1,2-氢转移反应，而Sn-β分子筛可以催化1,2-氢转移反应[111]，所以它在催化葡萄糖生成HMF[112]、乳酸[191,229]及其他α-羟基酸[245,271,272,312]中都具有较好的催化作用。因为$Ca(OH)_2$对催化1,2-氢转移反应和逆羟醛缩合反应都具有一定的催化作用，故它可用于转化葡萄糖生成乳酸和1,2-丙二醇[202,310]。因为WO_3对逆羟醛缩合反应具有催化作用，故含钨的催化剂在转化纤维素生成乙二醇和乳酸的研究中都表现出一定的活性[30,34,297]。此外，硝酸铅、硫酸氧钒、氯化铒、氯化锌及Sn-β分子筛等均能够催化葡萄糖转化为乳酸[30,191,214-218,258-260]，说明这些催化剂都能够催化逆羟醛缩合反应和 1,2-氢转移反应；如果将这些催化剂与加氢催化剂相结合并适当调整加氢活性，则有望开发出对葡萄糖及纤维素制备 1,2-丙二醇具有较高催化活性的催化剂体系；如果将这些催化剂与酸性催化剂相结合从而抑制其逆羟醛缩合反应，则有可能开发出对 HMF 具有高选择性的催化体系。

2.9 葡萄糖异构化机理

如前所述，葡萄糖转化为 HMF、乳酸和 1,2-丙二醇这三种重要的平台化学品都涉及葡萄糖异构化为果糖这一步骤。目前的研究发现，对葡萄糖异构化为果糖这一反应具有催化作用的催化剂可以分为三类：碱性催化剂［如 NaOH、$Ca(OH)_2$、$Ba(OH)_2$、镁铝水滑石等］、均相路易斯酸催化剂［如 $CrCl_3$、$AlCl_3$、$SnCl_4$、$Pb(NO_3)_2$ 等］和非均相路易斯酸催化剂（如 Sn-β、AlW 等）。由于葡萄糖在水溶液中主要以吡喃环形式存在，所以通常认为葡萄糖异构化为果糖涉及三个主要步骤：①吡喃环式葡萄糖开环成为链式葡萄糖；②链式葡萄糖在催化剂作用下经过 1,2-氢转移反应而转化为链式果糖；③链式果糖从催化剂脱离并闭环成为呋喃环式果糖[313]。上述三个步骤中，1,2-氢转移反应是整个过程的速控步[314]。对于三种不同的催化剂，目前认为其催化活性中心不同，所以人们提出了不同的催化作用机理，下面分别进行介绍。

2.9.1 碱催化葡萄糖异构化为果糖

Tessonnier 等提出的碱催化葡萄糖异构化机理如图 2-19 所示：①在 OH^- 作用下，吡喃环式葡萄糖的 C_1 原子❶上的氧原子带负电荷；②葡萄糖开环成为链式葡萄糖，而负电荷转移到 C_5 原子上的氧原子上；③C_2 原子的氢原子转移到 C_5 原子上的氧原子上形成羟基，而负电荷转移到 C_2 原子上，同时，C_1 原子上的氧原子与 C_2 原子羟基上的氢原子形成氢键，于是，在 C_1 原子和 C_2 原子之间形成烯醇式结构中间体；④C_2 原子羟基上的氢原子完全转移到 C_1 原子的氧原子上，从而将 C_1 原子上的羰基转变为羟基，而 C_2 原子上的羟基转变为羰基，从而成为带负电荷的链式果糖；⑤带负电荷的链式果糖缩醛环化为

---

❶ 指 1 位碳原子。

环式果糖，再与水分子结合释放出 OH$^-$[315,316]。但是，该路径不能说明 C_1 原子是如何增加一个氢原子的，也未能阐明负电荷是如何从 C_5 原子上的氧原子转移到 C_2 原子，又如何从 C_2 原子转移到 C_5 原子的氧原子上。因此，笔者认为该路径并不完善。

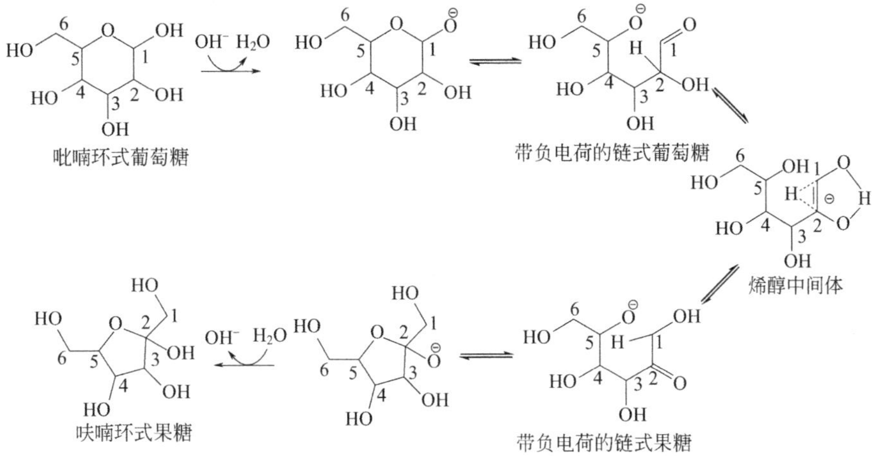

图 2-19 碱催化葡萄糖异构化机理

从电荷转移的角度考虑，笔者提出的碱催化链式葡萄糖异构化的反应机理如图 2-20 所示。碱催化剂中的 OH$^-$ 阴离子可能首先攻击带正电荷的 C_2 原子上的 α-H 原子形成水分子，并将负电荷转移到 C_2 原子上。随后，负电荷从 C_2 原子转移到 C_1 原子上，并在 C_2 原子和 C_1 原子之间形成 C═C 双键，也就是烯醇式中间体。与此同时，C_2 原子的羟基上的氢原子与 C_1 原子的氧原子形成氢键。由于负电荷存在于 C_1 原子上，其吸电子能力较强，导致 C_2 原子羟基中的氢原子转移到 C_1 原子上，从而使得 C_1 原子上的羰基转变为羟基，而 C_2 原子上的羟基失去氢原子变为羰基。最后，带负电荷的 C_1 原子与水分子相结合并夺走水中的氢原子到 C_1 原子上，并使得水分子转变为 OH$^-$，实现催化剂的再生。至于吡喃环式葡萄糖转化为链式葡萄糖及链式果糖转化为呋喃环式果糖，笔者认为都是一些较易发生的过程，在温度较高的条件下即使没有催化剂也能够顺利进行。

2.9.2 均相路易斯酸催化葡萄糖异构化的机理

研究发现，均相路易斯酸（CrCl$_3$、AlCl$_3$）在水溶液中会与水分子相结合并发生部分水解，分别形成活性物种 [Cr(H$_2$O)$_5$OH]$^{2+}$ 和 [Al(H$_2$O)$_5$OH]$^{2+}$

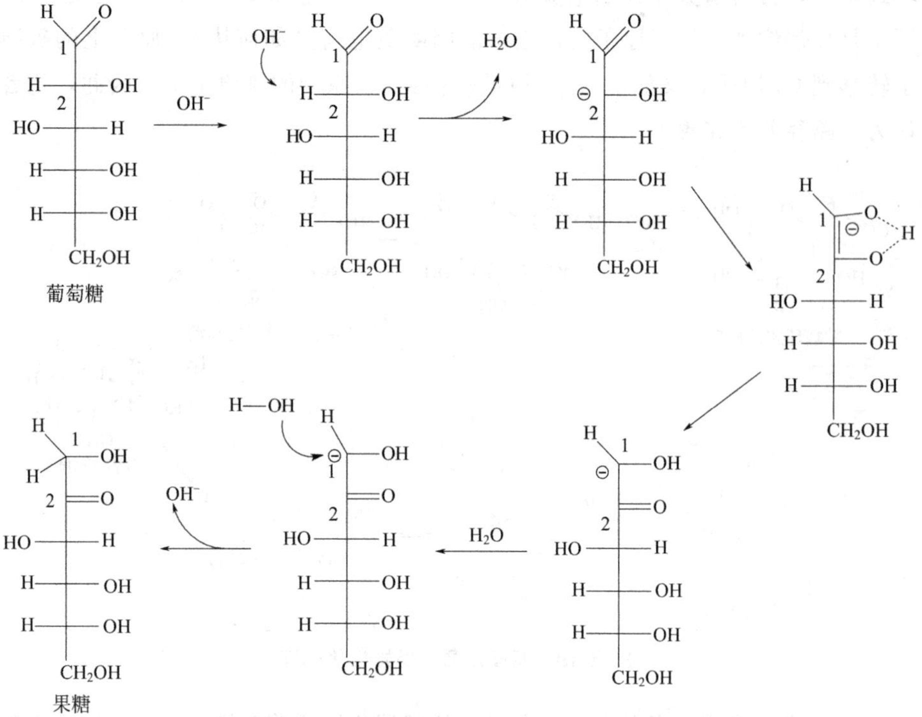

图 2-20　笔者提出的碱催化链式葡萄糖异构化的反应机理

(结构见图 2-21)[159,314]。该类活性中心在水溶液中呈八面体结构，其中，金属阳离子中心是路易斯酸性位，而与金属阳离子结合的碱性基团 OH⁻ 起到布朗斯特碱的作用。

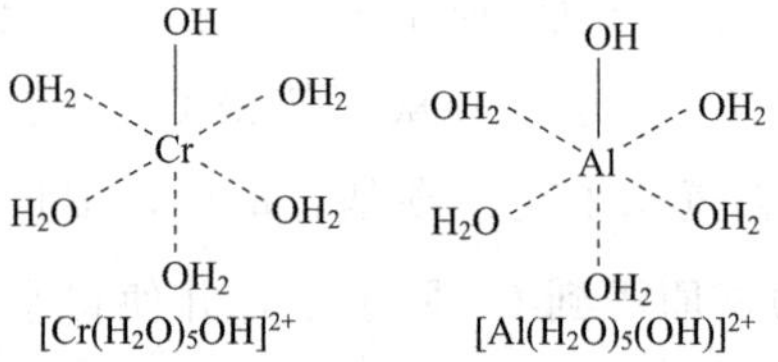

图 2-21　Cr^{3+}、Al^{3+} 在水溶液中形成的催化活性物种

$CrCl_3$ 对葡萄糖异构化反应的催化机理如图 2-22 所示[65,159,314,317]：① 葡萄糖的羰基（C=O）作为电子供体在金属离子（路易斯酸中心）发生吸附，葡萄糖上的 C_1-O❶ 和 C_2-O 取代催化剂金属阳离子上的两个水分子并与之配位，形成 $[Cr(C_6H_{12}O_6)(H_2O)_3OH]^{2+}$ 阳离子，同时 C_2-OH 与 Cr-OH 的碱性位点形成氢

---

❶ 指与 1 位碳原子相连的氧原子，C_2-O 指与 2 位碳原子相连的氧原子，以此类推。

键；②在 $[Cr(C_6H_{12}O_6)(H_2O)_3OH]^{2+}$ 阳离子中，葡萄糖 C_2-OH 上的氢原子转移到催化剂的 Cr-OH 上，形成 C=O 和 Cr-OH$_2$，失去质子的葡萄糖带负电荷，络合阳离子转变为 $[Cr(C_6H_{11}O_6)(H_2O)_4]^{2+}$；③$[Cr(C_6H_{11}O_6)(H_2O)_4]^{2+}$ 阳离子中，失去质子的葡萄糖 C_2-O 原子上的负电荷转移到 C_1-O 原子上，并导致 C_2 原子上的氢原子向 C_1 原子转移；④由于 C_1-O 原子带负点，$[Cr(C_6H_{11}O_6)(H_2O)_4]^{2+}$ 阳离子中参与配位的水分子中心（路易斯酸性位点）的一个氢原子转移到 C_1-O 上去，实现 C=O 转变为 C_1-OH 和 Cr-OH 的再生，从而得到链式果糖参与配位的 $[Cr(C_6H_{12}O_6)(H_2O)_3OH]^{2+}$ 阳离子；⑤链式果糖从 $[Cr(C_6H_{12}O_6)(H_2O)_3OH]^{2+}$ 阳离子中脱离。在整个过程中，催化剂的活性位点由葡萄糖分子内 C_2 原子上的氢原子转移到 C_1 原子上是整个反应的速控步[314]。

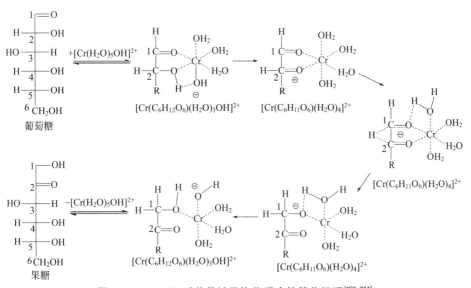

图 2-22 $CrCl_3$ 对葡萄糖异构化反应的催化机理[159, 314]

Tang 等分析 $AlCl_3$ 催化葡萄糖异构化反应，利用 ESI-MS/MS 证明 $AlCl_3$ 在水溶液中的活性物种为 $[Al(H_2O)_4(OH)_2]^+$ 而不是 $[Al(H_2O)_5(OH)]^{2+}$。另外，他们认为葡萄糖上的 C_1-O 与金属阳离子活性中心的相互作用对反应的触发起重要作用[318]。

铬盐在离子液体中催化葡萄糖异构化的机理与在水中催化葡萄糖异构化的机理相似，不过，在离子液体中，Cr^{3+} 与 Cl^- 和葡萄糖的羟基发生配位，双核 Cr 配合物稳定了链式葡萄糖异构化为链式果糖的中间体，而且链式葡萄糖发生 1,2-氢转移是在双核 Cr 配合物的作用下进行的[313]。

如前面所述，含 Zn^{2+}、Pb^{2+}、VO^{2+}、Er^{2+} 的催化剂均能高效催化葡萄糖生成乳酸，表明这几种金属阳离子都能催化葡萄糖异构化为果糖。关于这几种阳离子在水溶液中所对应的活性物种及其催化机理，目前并未有文献进行报道。不过，极有可能这几种金属阳离子在水溶液中也是形成了活性物种 $[M(H_2O)_x(OH)_y]^{z+}$，该活性物种通过与 Cr^{3+} 和 Al^{3+} 相似的催化机理实现葡萄糖和果糖的异构化反应[317]。

2.9.3 非均相路易斯酸 Sn-β 催化葡萄糖异构化的机理

由于 Sn-β 对葡萄糖的异构化反应具有优异的催化效果，大量工作研究了 Sn-β 对葡萄糖异构化反应的催化作用机理。在早期的研究中，普遍认为 Sn-OH 是 Sn-β 的催化活性中心，而与 Sn^{4+} 中心相邻的硅醇未起到催化作用（活性中心见图 2-23）。然而，Bermejo-Deval 等利用 Na^+ 与 Sn-β 中硅醇上的氢离子交换从而去除硅醇结构，得到的 Na-Sn-β 不能很好地催化葡萄糖的异构化反应，表明 Sn-β 中硅醇基团 Si-OH 对葡萄糖的异构化反应起到重要作用[319]。Li 等通过周期性密度泛函计算，指出 Sn-β 分子筛中与 Sn^{4+} 相邻的羟基（可能来自硅醇结构，也可能来自吸附于 Sn-β 上的水分子）能够降低葡萄糖异构化反应过程中形成过渡态的活化能，对于葡萄糖的异构化反应具有重要的影响[320]。Rai 等同样在密度泛函计算的基础上，认为 Sn-β 催化葡萄糖异构化为果糖时，葡萄糖并不是与 Sn^{4+} 形成双齿配位，而是由 C_2-O 与 Sn^{4+} 配位，C_1-O 则与和 Sn^{4+} 相邻的 Si-OH 通过氢键相连[321]。基于这些后期的发现，部分研究人员提出了 Sn-β 固体酸中 Sn-OH 和与之相邻的 Si-OH 是催化葡萄糖异构化的活性中心（如图 2-23 中有硅醇参与的活性中心结构）。

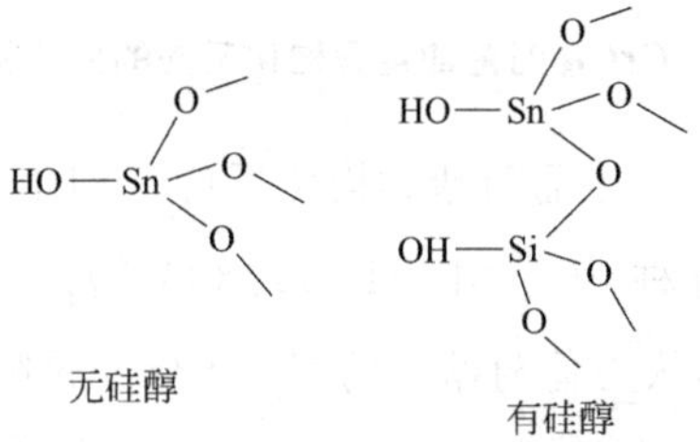

无硅醇　　　　有硅醇

图 2-23　固体酸 Sn-β 的催化活性中心

早期提出的无硅醇参与的 Sn-β 催化葡萄糖异构化反应的机理如图 2-24 所示[314,322-324]：①葡萄糖的 C_2-O 和 C_1-O 与路易斯酸中心 Sn^{4+} 形成双齿配位，同时 Sn-OH 与葡萄糖的 C_2-OH 之间形成氢键，得到过渡态 1；②Sn-OH 从 C_2-OH 中夺取氢原子，形成 Sn-OH_2 和 C=O，C_2 上的氢原子逐渐向 C_1 转移，得到过渡态 2；③Sn-OH_2 将氢原子传递给 C_1-O，促使 C_2 上的氢原子转移到 C_1 上，从而生成链式果糖分子；④链式果糖分子从 Sn^{4+} 配位中心脱离，形成链式果糖。在这种双齿配位机理中，Sn^{4+} 与均相催化过程中的 Cr^{3+}、Al^{3+} 具有相似的作用。

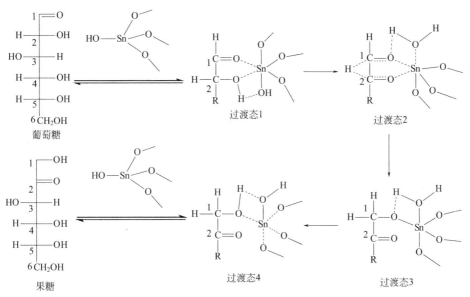

图 2-24　无硅醇参与的 Sn-β 催化葡萄糖异构化反应的机理[322]

硅醇参与的 Sn-β 催化葡萄糖异构化反应的机理如图 2-25 所示：①葡萄糖中 C_1 位上的羰基与硅醇 Si-OH 形成氢键，而 C_2-OH 中的氧原子与锡原子配合成键，同时氢原子则与 Sn-OH 形成氢键，得到过渡态 1；②C_2-OH 中的氢原子转移到 Sn-OH 上去形成 Sn-OH_2，同时 C_1-O 中的氧原子与 Sn^{4+} 成键并导致 C_1=O 的双键断裂，C_2 上的氢原子开始向 C_1 转移，形成过渡态 2；③C_2-O 从 Sn^{4+} 上脱离，形成 C_2=O，C_2 上的氢原子则完全转移到 C_1 上去，形成过渡态 3；④硅醇上的氢原子转移到 C_1-O 上去，形成过渡态 4，最后从 Sn-β 上脱离，形成链式果糖分子。

目前关于葡萄糖异构化为果糖虽然提出了多种催化机理，但是这些机理都还不能很好地解释这些催化剂对葡萄糖异构化反应的催化作用，也不能够指导固体催化剂的研发。因此，关于葡萄糖异构化为果糖的催化作用机理仍然需要更多的研究。

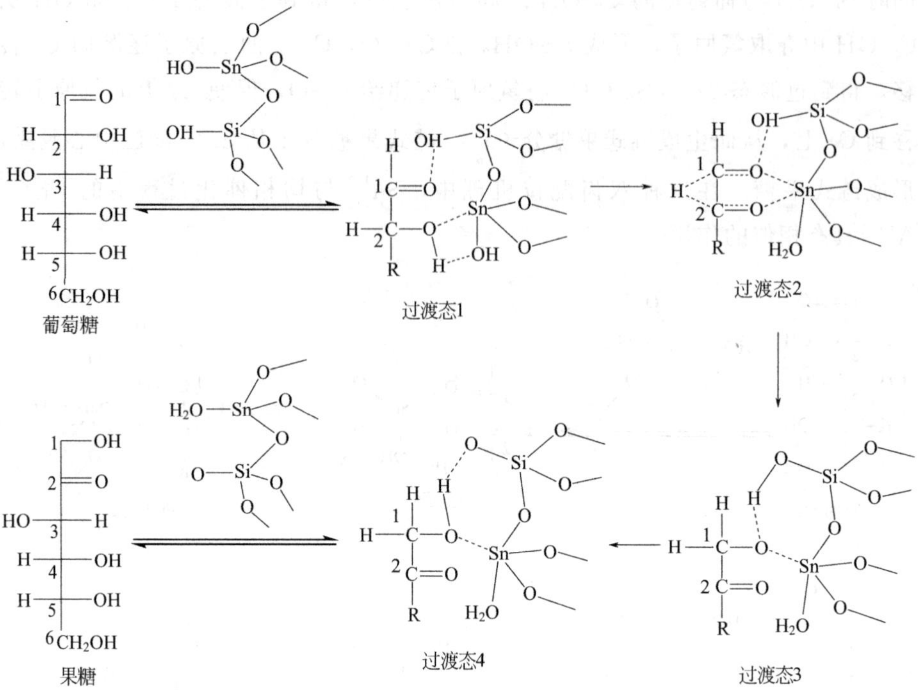

图 2-25 硅醇参与的 Sn-β 催化葡萄糖异构化反应的机理

参 考 文 献

[1] Galkin K I, Ananikov V P. When will 5-hydroxymethylfurfural, the "Sleeping Giant" of sustainable chemistry, awaken? ChemSusChem, 2019, 12: 2976-2982.

[2] De Clercq R, Dusselier M, Sels B F. Heterogeneous catalysis for bio-based polyester monomers from cellulosic biomass: advances, challenges and prospects. Green Chemistry, 2017, 19 (21): 5012-5040.

[3] Zhang Z, Song J, Han B. Catalytic transformation of lignocellulose into chemicals and fuel products in ionic liquids. Chemical Reviews, 2017, 117 (10): 6834-6880.

[4] Corma A, Iborra S, Velty A. Chemical routes for the transformation of biomass into chemicals. Chemical Reviews, 2007, 107 (6): 2411-2502.

[5] Huber G W, Iborra S, Corma A. Synthesis of transportation fuels from biomass: chemistry, catalysts, and engineering. Chemical Reviews, 2006, 106 (9): 4044-4098.

[6] Hu L, Lin L, Wu Z, Zhou S, Liu S. Chemocatalytic hydrolysis of cellulose into glucose over solid acid catalysts. Applied Catalysis B: Environmental, 2015, 174-175: 225-243.

[7] Rinaldi R, Schuth F. Acid hydrolysis of cellulose as the entry point into biorefinery schemes. ChemSusChem, 2009, 2 (12): 1096-1107.

[8] Huang Y B, Fu Y. Hydrolysis of cellulose to glucose by solid acid catalysts. Green Chemistry,

2013, 15 (5): 1095-1111.

[9] Yu Y, Lou X, Wu H. Some recent advances in hydrolysis of biomass in hot-compressed water and its comparisons with other hydrolysis methods. Energy and Fuels, 2008, 22 (1): 46-60.

[10] Yu Y, Wu H. Significant differences in the hydrolysis behavior of amorphous and crystalline portions within microcrystalline cellulose in hot-compressed water, Industrial and Engineering Chemistry Research. 2010.

[11] Peterson A A, Vogel F, Lachance R P, Fröling M, Antal J M J, Tester J W. Thermochemical biofuel production in hydrothermal media: a review of sub-and supercritical water technologies. Energy and Environmental Science, 2008, 1 (1): 32-65.

[12] Deguchi S, Tsujii K, Horikoshi K. Cooking cellulose in hot and compressed water. Chemical Communications, 2006, (31): 3293-3295.

[13] Kang S M, Li X H, Fan J, Chang J. Characterization of hydrochars produced by hydrothermal carbonization of lignin, cellulose, D-xylose and wood meal. Industrial and Engineering Chemistry Research, 2012, 51 (26): 9023-9031.

[14] Sevilla M, Fuertes A B. The production of carbon materials by hydrothermal carbonization of cellulose. Carbon, 2009, 47 (9): 2281-2289.

[15] Lu X W, Pellechia P J, Flora J R V, Berge N D. Influence of reaction time and temperature on product formation and characteristics associated with the hydrothermal carbonization of cellulose. Bioresource Technology, 2013, 138: 180-190.

[16] Akhtar J, Amin N A S. A review on process conditions for optimum bio-oil yield in hydrothermal liquefaction of biomass. Renewable and Sustainable Energy Reviews, 2011, 15 (3): 1615-1624.

[17] Yin S D, Tan Z C. Hydrothermal liquefaction of cellulose to bio-oil under acidic, neutral and alkaline conditions. Applied Energy, 2012, 92: 234-239.

[18] Liu Y, Luo C, Liu H C. Tungsten trioxide promoted selective conversion of cellulose into propylene glycol and ethylene glycol on a ruthenium catalyst. Angewandte Chemie-International Edition, 2012, 51 (13): 3249-3253.

[19] Peng L C, Lin L, Zhang J H, Zhuang J P, Zhang B X, Gong Y. Catalytic conversion of cellulose to levulinic acid by metal chlorides. Molecules, 2010, 15 (8): 5258-5272.

[20] Shi N, Liu Q Y, Zhang Q, Wang T J, Ma L L. High yield production of 5-hydroxymethylfurfural from cellulose by high concentration of sulfates in biphasic system. Green Chemistry, 2013, 15 (7): 1967-1974.

[21] Suganuma S, Nakajima K, Kitano M, Yamaguchi D, Kato H, Hayashi S, Hara M. Hydrolysis of cellulose by amorphous carbon bearing SO_3H, COOH, and OH groups. Journal of the American Chemical Society, 2008, 130 (38): 12787-12793.

[22] Onda A, Ochi T, Yanagisawa K. Selective hydrolysis of cellulose into glucose over solid acid catalysts. Green Chemistry, 2008, 10 (10): 1033-1037.

[23] Pang J F, Wang A Q, Zheng M Y, Zhang T. Hydrolysis of cellulose into glucose over carbons sulfonated at elevated temperatures. Chemical Communications, 2010, 46 (37): 6935-6937.

[24] Wu Y, Fu Z, Yin D, Xu Q, Liu F, Lu C, Mao L. Microwave-assisted hydrolysis of crystalline

cellulose catalyzed by biomass char sulfonic acids. Green Chemistry, 2010, 12 (4): 696-700.

[25] Yamaguchi D, Kitano M, Suganuma S, Nakajima K, Kato H, Hara M. Hydrolysis of cellulose by a solid acid catalyst under optimal reaction conditions. Journal of Physical Chemistry C. 2009, 113 (8): 3181-3188.

[26] Van De Vyver S, Peng L, Geboers J, Schepers H, De Clippel F, Gommes C J, Goderis B, Jacobs P A, Sels B F. Sulfonated silica/carbon nanocomposites as novel catalysts for hydrolysis of cellulose to glucose. Green Chemistry, 2010, 12 (9): 1560.

[27] Guo H, Qi X, Li L, Smith R L, Jr. Hydrolysis of cellulose over functionalized glucose-derived carbon catalyst in ionic liquid. Bioresource Technology, 2012, 116: 355-359.

[28] Rinaldi R, Palkovits R, Schuth F. Depolymerization of cellulose using solid catalysts in ionic liquids. Angewandte Chemie-International Edition, 2008, 47 (42): 8047-8050.

[29] Shuai L, Pan X. Hydrolysis of cellulose by cellulase-mimetic solid catalyst. Energy and Environmental Science, 2012, 5 (5): 6889.

[30] Chambon F, Rataboul F, Pinel C, Cabiac A, Guillon E, Essayem N. Cellulose hydrothermal conversion promoted by heterogeneous Brønsted and Lewis acids: Remarkable efficiency of solid Lewis acids to produce lactic acid. Applied Catalysis B-Environmental. 2011, 105 (1-2): 171-181.

[31] Gliozzi G, Innorta A, Mancini A, Bortolo R, Perego C, Ricci M, Cavani F. Zr/P/O catalyst for the direct acid chemo-hydrolysis of non-pretreated microcrystalline cellulose and softwood sawdust. Applied Catalysis B: Environmental. 2014, 145: 24-33.

[32] Takagaki A, Tagusagawa C, Domen K. Glucose production from saccharides using layered transition metal oxide and exfoliated nanosheets as a water-tolerant solid acid catalyst. Chemical Communications, 2008, 14 (42): 5363-5365.

[33] Xi J, Ding D, Shao Y, Liu X, Lu G, Wang Y. Production of ethylene glycol and its monoether derivative from cellulose. Acs Sustainable Chemistry and Engineering, 2014, 2 (10): 2355-2362.

[34] Chambon F, Rataboul F, Pinel C, Cabiac A, Guillon E, Essayem N. Cellulose conversion with tungstated-alumina-based catalysts: influence of the presence of platinum and mechanistic studies. ChemSusChem. 2013, 6 (3): 500-507.

[35] Hamdy M S, Eissa M A, Keshk S M a S. New catalyst with multiple active sites for selective hydrogenolysis of cellulose to ethylene glycol. Green Chemistry, 2017, 19 (21): 5144-5151.

[36] Lai D M, Deng L, Li J A, Liao B, Guo Q X, Fu Y. Hydrolysis of Cellulose into Glucose by Magnetic Solid Acid. ChemSusChem. 2011, 4 (1): 55-58.

[37] Lai D M, Deng L, Guo Q X, Fu Y. Hydrolysis of biomass by magnetic solid acid. Energy and Environmental Science, 2011, 4 (9): 3552.

[38] Zhang C, Wang H, Liu F, Wang L, He H. Magnetic core-shell Fe_3O_4@$C-SO_3H$ nanoparticle catalyst for hydrolysis of cellulose. Cellulose, 2013, 20 (1): 127-134.

[39] Van Putten R J, Van Der Waal J C, De Jong E, Rasrendra C B, Heeres H J, De Vries J G. Hydroxymethylfurfural, a versatile platform chemical made from renewable resources. Chemical Reviews, 2013, 113 (3): 1499-1597.

[40] Rosatella A A, Simeonov S P, Frade R F M, Afonso C A M. 5-Hydroxymethylfurfural (HMF)

[41] Moreau C, Belgacem M N, Gandini A. Recent catalytic advances in the chemistry of substituted furans from carbohydrates and in the ensuing polymers. Topics in Catalysis, 2004, 27 (1-4): 11-30.

[42] Girisuta B, Janssen L P B M, Heeres H J. A kinetic study on the decomposition of 5-hydroxymethylfurfural into levulinic acid. Green Chemistry, 2006, 8 (8): 701-709.

[43] Mascal M, Nikitin E B. Direct, high-yield conversion of cellulose into biofuel. Angewandte Chemie-International Edition, 2008, 47 (41): 7924-7926.

[44] Balakrishnan M, Sacia E R, Bell A T. Etherification and reductive etherification of 5-(hydroxymethyl) furfural: 5-(alkoxymethyl) furfurals and 2, 5-bis (alkoxymethyl) furans as potential bio-diesel candidates. Green Chemistry, 2012, 14 (6): 1626-1634.

[45] Lanzafame P, Temi D M, Perathoner S, Spadaro A N, Centi G. Direct conversion of cellulose to glucose and valuable intermediates in mild reaction conditions over solid acid catalysts. Catalysis Today, 2012, 179 (1): 178-184.

[46] Dumesic J A, Roman-Leshkov Y, Barrett C J, Liu Z Y. Production of dimethylfuran for liquid fuels from biomass-derived carbohydrates. Nature, 2007, 447 (7147): 982-U985.

[47] Verdeguer P, Merat N, Gaset A. Catalytic-oxidation of HMF to 2, 5-furandicarboxylic acid. Journal of Molecular Catalysis, 1993, 85 (3): 327-344.

[48] Hopkins K T, Wilson W D, Bender B C, Mccurdy D R, Hall J E, Tidwell R R, Kumar A, Bajic M, Boykin D W. Extended aromatic furan amidino derivatives as anti-pneumocystis carinii agents. Journal of Medicinal Chemistry, 1998, 41 (20): 3872-3878.

[49] Gandini A, Coelho D, Gomes M, Reis B, Silvestre A. Materials from renewable resources based on furan monomers and furan chemistry: work in progress. Journal of Materials Chemistry, 2009, 19 (45): 8656-8664.

[50] Dumesic J A, Huber G W, Chheda J N, Barrett C J. Production of liquid alkanes by aqueous-phase processing of biomass-derived carbohydrates. Science, 2005, 308 (5727): 1446-1450.

[51] Chheda J N, Huber G W, Dumesic J A. Liquid-phase catalytic processing of biomass-derived oxygenated hydrocarbons to fuels and chemicals. Angewandte Chemie-International Edition, 2007, 46 (38): 7164-7183.

[52] Chatterjee M, Matsushima K, Ikushima Y, Sato M, Yokoyama T, Kawanami H, Suzuki T. Production of linear alkane via hydrogenative ring opening of a furfural-derived compound in supercritical carbon dioxide. Green Chemistry, 2010, 12 (5): 779-782.

[53] Chheda J N, Dumesic J A. An overview of dehydration, aldol-condensation and hydrogenation processes for production of liquid alkanes from biomass-derived carbohydrates. Catalysis Today, 2007, 123 (1-4): 59-70.

[54] Dedsuksophon W, Faungnawakij K, Champreda V, Laosiripojana N. Hydrolysis/dehydration/aldol-condensation/hydrogenation of lignocellulosic biomass and biomass-derived carbohydrates in the presence of Pd/WO_3-ZrO_2 in a single reactor. Bioresource Technology, 2011, 102 (2): 2040-2046.

[55] Xu W, Liu X, Ren J, Zhang P, Wang Y, Guo Y, Guo Y, Lu G. A novel mesoporous Pd _ cobalt aluminate bifunctional catalyst for aldol condensation and following hydrogenation. Catalysis Communications, 2010, 11 (8): 721-726.

[56] Li X, Xu R, Yang J, Nie S, Liu D, Liu Y, Si C. Production of 5-hydroxymethylfurfural and levulinic acid from lignocellulosic biomass and catalytic upgradation. Industrial Crops and Products, 2019, 130: 184-197.

[57] Wang T F, Nolte M W, Shanks B H. Catalytic dehydration of C-6 carbohydrates for the production of hydroxymethylfurfural (HMF) as a versatile platform chemical. Green Chemistry, 2014, 16 (2): 548-572.

[58] Dutta S, De S, Saha B. Advances in biomass transformation to 5-hydroxymethylfurfural and mechanistic aspects. Biomass and Bioenergy, 2013, 55: 355-369.

[59] Lewkowski J. Synthesis, chemistry and applications of 5-hydroxymethylfurfural and its derivatives. Cheminform, 2001, 34 (2): 17-54.

[60] James O O, Maity S, Usman L A, Ajanaku K O, Ajani O O, Siyanbola T O, Sahu S, Chaubey R. Towards the conversion of carbohydrate biomass feedstocks to biofuels via hydroxylmethylfurfural. Energy and Environmental Science, 2010, 3 (12): 1833-1850.

[61] Kuster B F M. 5-hydroxymethylfurfural (HMF)-a review focusing on its manufacture. Starch-Starke, 1990, 42 (8): 314-321.

[62] Saha B, Abu-Omar M M. Advances in 5-hydroxymethylfurfural production from biomass in biphasic solvents. Green Chemistry, 2014, 16 (1): 24-38.

[63] Dutta S, Pal S. Promises in direct conversion of cellulose and lignocellulosic biomass to chemicals and fuels: Combined solvent-nanocatalysis approach for biorefinary. Biomass and Bioenergy, 2014, 62: 182-197.

[64] Jadhav H, Pedersen C M, Solling T, Bols M. 3-Deoxy-glucosone is an intermediate in the formation of furfurals from D-glucose. ChemSusChem, 2011, 4 (8): 1049-1051.

[65] Mushrif S H, Varghese J J, Vlachos D G. Insights into the Cr (III) catalyzed isomerization mechanism of glucose to fructose in the presence of water using ab initio molecular dynamics. Physical Chemistry Chemical Physics, 2014, 16 (36): 19564-19572.

[66] Pagan-Torres Y J, Wang T F, Gallo J M R, Shanks B H, Dumesic J A. Production of 5-hydroxymethylfurfural from glucose using a combination of lewis and brønsted Acid Catalysts in Water in a Biphasic Reactor with an Alkylphenol Solvent. Acs Catalysis, 2012, 2 (6): 930-934.

[67] Shi N, Liu Q, He X, Wang G, Chen N, Peng J, Ma L. Molecular structure and formation mechanism of hydrochar from hydrothermal carbonization of carbohydrates. Energy and Fuels, 2019, 33 (10): 9904-9915.

[68] Shi N, Liu Q Y, Ju R M, He X, Zhang Y L, Tang S Y, Ma L L. Condensation of α-carbonyl aldehydes leads to the formation of solid humins during the hydrothermal degradation of carbohydrates. ACS Omega, 2019, 4 (4): 7330-7343.

[69] Anet E F L J. 3-deoxyglycosuloses (3-deoxyglycosones) and the degradation of carbohydrates. Advances in Carbohydrate Chemistry, 1964, 19: 181-218.

[70] Asghari F S, Yoshida H. Conversion of Japanese red pine wood (Pinus densiflora) into valuable chemicals under subcritical water conditions. Carbohydrate Research, 2010, 345 (1): 124-131.

[71] Shi N, Liu Q Y, Ma L L, Wang T J, Zhang Q, Zhang Q, Liao Y H. Direct degradation of cellulose to 5-hydroxymethylfurfural in hot compressed steam with inorganic acidic salts. Rsc Advances, 2014, 4 (10): 4978-4984.

[72] Shi N, Liu Q Y, Wang T J, Ma L L, Zhang Q, Zhang Q. One-pot degradation of cellulose into furfural compounds in hot compressed steam with dihydric phosphates. Acs Sustainable Chemistry and Engineering, 2014, 2 (4): 637-642.

[73] Corma A, Iborra S, Velty A. Chemical routes for the transformation of biomass into chemicals. Chemical Reviews, 2007, 107: 2411-2502.

[74] Binder J B, Raines R T. Simple chemical transformation of lignocellulosic biomass into furans for fuels and chemicals. Journal of the American Chemical Society, 2009, 131 (5): 1979-1985.

[75] Zakrzewska M E, Bogel-Lukasik E, Bogel-Lukasik R. Ionic Liquid-Mediated Formation of 5-Hydroxymethylfurfural-A Promising Biomass-Derived Building Block. Chemical Reviews, 2011, 111 (2): 397-417.

[76] Moreau C, Finiels A, Vanoye L. Dehydration of fructose and sucrose into 5-hydroxymethylfurfural in the presence of 1-H-3-methyl imidazolium chloride acting both as solvent and catalyst. Journal of Molecular Catalysis A: Chemical, 2006, 253 (1-2): 165-169.

[77] Zhao H B, Holladay J E, Brown H, Zhang Z C. Metal chlorides in ionic liquid solvents convert sugars to 5-hydroxymethylfurfural. Science, 2007, 316 (5831): 1597-1600.

[78] Antal M J, Mok W S L, Richards G N. Kinetic-studies of the reactions of ketoses and aldoses in water at high-temperature. 1. mechanism of formation of 5-(hydroxymethyl)-2-furaldehyde from D-fructose and sucrose. Carbohydrate Research, 1990, 199 (1): 91-109.

[79] Rigal L, Gaset A, Gorrichon J P. Selective conversion of D-fructose to 5-hydroxymethyl-2-furancarboxaldehyde using a water-solvent-ion-exchange resin triphasic System. Industrial and Engineering Chemistry Product Research and Development, 1981, 20 (4): 719-721.

[80] Dam H E V, Kieboom A P G, Bekkum H V. The conversion of fructose and glucose in acidic media: formation of hydroxymethylfurfural. Starch-Strke, 1986, 38 (3): 95-101.

[81] Widsten P, Murton K, West M. Production of 5-hydroxymethylfurfural and furfural from a mixed saccharide feedstock in biphasic solvent systems. Industrial Crops and Products, 2018, 119: 237-242.

[82] Roman-Leshkov Y, Dumesic J A. Solvent effects on fructose dehydration to 5-hydroxymethylfurfural in biphasic systems saturated with inorganic salts. Topics in Catalysis, 2009, 52 (3): 297-303.

[83] Roman-Leshkov Y, Chheda J N, Dumesic J A. Phase modifiers promote efficient production of hydroxymethylfurfural from fructose. Science, 2006, 312 (5782): 1933-1937.

[84] Li C Z, Zhao Z K, Wang A Q, Zheng M Y, Zhang T. Production of 5-hydroxymethylfurfural in ionic liquids under high fructose concentration conditions. Carbohydrate Research, 2010, 345 (13): 1846-1850.

[85] Lai L K, Zhang Y G. The Production of 5-hydroxymethylfurfural from fructose in isopropyl alcohol: a green and efficient system. ChemSusChem, 2011, 4 (12): 1745-1748.

[86] Chheda J N, Roman-Leshkov Y, Dumesic J A. Production of 5-hydroxymethylfurfural and furfural by dehydration of biomass-derived mono-and poly-saccharides. Green Chemistry, 2007, 9 (4): 342-350.

[87] Caes B R, Raines R T. Conversion of fructose into 5-(hydroxymethyl) furfural in Sulfolane. ChemSusChem, 2011, 4 (3): 353-356.

[88] Cao Q, Guo X C, Yao S X, Guan J, Wang X Y, Mu X D, Zhang D K. Conversion of hexose into 5-hydroxymethylfurfural in imidazolium ionic liquids with and without a catalyst. Carbohydrate Research, 2011, 346 (7): 956-959.

[89] Yong G, Zhang Y G, Ying J Y. Efficient catalytic system for the selective production of 5-hhydroxymethylfurfural from glucose and fructose. Angewandte Chemie-International Edition, 2008, 47 (48): 9345-9348.

[90] Zhang Z H, Zhao Z B K. Microwave-assisted conversion of lignocellulosic biomass into furans in ionic liquid. Bioresource Technology, 2010, 101 (3): 1111-1114.

[91] Kim B, Jeong J, Lee D, Kim S, Yoon H J, Lee Y S, Cho J K. Direct transformation of cellulose into 5-hydroxymethyl-2-furfural using a combination of metal chlorides in imidazolium ionic liquid. Green Chemistry, 2011, 13 (6): 1503-1506.

[92] Qi X H, Watanabe M, Aida T M, Smith R L. Catalytic conversion of cellulose into 5-hydroxymethylfurfural in high yields via a two-step process. Cellulose, 2011, 18 (5): 1327-1333.

[93] Yu H B, Wang P, Zhan S H, Wang S Q. Catalytic hydrolysis of lignocellulosic biomass into 5-hydroxymethylfurfural in ionic liquid. Bioresource Technology. 2011, 102 (5): 4179-4183.

[94] Chen T M, Lin L. Conversion of glucose in CPL-LiCl to 5-hydroxymethylfurfural. Research Progress in Paper Industry and Biorefinery, 2010, 1-3: 1421-1424.

[95] Bond J Q, Alonso D M, Wang D, West R M, Dumesic J A. Integrated catalytic conversion of γ-valerolactone to liquid alkenes for transportation fuels. Science, 2010, 327 (5969): 1110-1114.

[96] Xie H B, Zhang Z H, Wang Q A, Liu W J, Zhao Z B. Catalytic conversion of carbohydrates into 5-hydroxymethylfurfural by germanium (IV) chloride in ionic liquids. ChemSusChem, 2011, 4 (1): 131-138.

[97] Yang Y, Hu C W, Abu-Omar M M. Conversion of carbohydrates and lignocellulosic biomass into 5-hydroxymethylfurfural using $AlCl_3 \cdot 6H_2O$ catalyst in a biphasic solvent system. Green Chemistry, 2012, 14 (2): 509-513.

[98] Hu S Q, Zhang Z F, Song J L, Zhou Y X, Han B X. Efficient conversion of glucose into 5-hydroxymethylfurfural catalyzed by a common Lewis acid $SnCl_4$ in an ionic liquid. Green Chemistry, 2009, 11 (11): 1746-1749.

[99] Davis M E, Nikolla E N, E., Roman-Leshkov Y, Moliner M. "One-pot" synthesis of 5-(hydroxymethyl) furfural from carbohydrates using Tin-Beta Zeolite. Acs Catalysis, 2011, 1 (4): 408-410.

[100] Chung C H, Chun J A, Lee J W, Yi Y B, Hong S S. Catalytic production of hydroxymethylfur-

[101] Azadi P, Carrasquillo-Flores R, Pagan-Torres Y J, Gurbuz E I, Farnood R, Dumesic J A. Catalytic conversion of biomass using solvents derived from lignin. Green Chemistry, 2012, 14 (6): 1573-1576.

[102] Qi X H, Watanabe M, Aida T M, Smith R L. Catalytical conversion of fructose and glucose into 5-hydroxymethylfurfural in hot compressed water by microwave heating. Catalysis Communications, 2008, 9 (13): 2244-2249.

[103] Dutta S, De S, Patra A K, Sasidharan M, Bhaumik A, Saha B. Microwave assisted rapid conversion of carbohydrates into 5-hydroxymethylfurfural catalyzed by mesoporous TiO_2 nanoparticles. Applied Catalysis A: General, 2011, 409: 133-139.

[104] Qi X H, Watanabe M, Aida T M, Smith R L. Synergistic conversion of glucose into 5-hydroxymethylfurfural in ionic liquid-water mixtures. Bioresource Technology, 2012, 109: 224-228.

[105] Yan H P, Yang Y, Tong D M, Xiang X, Hu C W. Catalytic conversion of glucose to 5-hydroxymethylfurfural over SO_4^{2-}/ZrO_2 and SO_4^{2-}/ZrO_2-Al_2O_3 solid acid catalysts. Catalysis Communications, 2009, 10 (11): 1558-1563.

[106] Hara M, Nakajima K, Baba Y, Noma R, Kitano M, Kondo J N, Hayashi S. $Nb_2O_5 \cdot nH_2O$ as a heterogeneous catalyst with water-tolerant Lewis acid sites. Journal of the American Chemical Society, 2011, 133 (12): 4224-4227.

[107] Du Y G, Yang F L, Liu Q S, Yue M, Bai X F. Tantalum compounds as heterogeneous catalysts for saccharide dehydration to 5-hydroxymethylfurfural. Chemical Communications. 2011, 47 (15): 4469-4471.

[108] Jadhav H, Taarning E, Pedersen C M, Bols M. Conversion of D-glucose into 5-hydroxymethylfurfural (HMF) using zeolite in [Bmim] Cl or tetrabutylammonium chloride (TBAC)/$CrCl_2$. Tetrahedron Letters, 2012, 53 (8): 983-985.

[109] Wang X H, Fan C Y, Guan H Y, Zhang H, Wang J H, Wang S T. Conversion of fructose and glucose into 5-hydroxymethylfurfural catalyzed by a solid heteropolyacid salt. Biomass and Bioenergy, 2011, 35 (7): 2659-2665.

[110] Guo F, Fang Z, Zhou T J. Conversion of fructose and glucose into 5-hydroxymethylfurfural with lignin-derived carbonaceous catalyst under microwave irradiation in dimethyl sulfoxide-ionic liquid mixtures. Bioresource Technology, 2012, 112 313-318.

[111] Davis M E, Moliner M, Roman-Leshkov Y. Tin-containing zeolites are highly active catalysts for the isomerization of glucose in water. Proceedings of the National Academy of Sciences of the United States of America, 2010, 107 (14): 6164-6168.

[112] Nikolla E, Roman-Leshkov Y, Moliner M, Davis M E. "One-pot" synthesis of 5-(hydroxymethyl) furfural from carbohydrates using Tin-beta zeolite. Acs Catalysis, 2011, 1 (4): 408-410.

[113] Himmel M E, Ding S Y, Johnson D K, Adney W S, Nimlos M R, Brady J W, Foust T D. Biomass recalcitrance: engineering plants and enzymes for biofuels production (vol 315, pg 804, 2007). Science, 2007, 315 (5813): 804-807.

[114] Zhao X B, Zhang L H, Liu D H. Biomass recalcitrance. Part II: Fundamentals of different pretreatments to increase the enzymatic digestibility of lignocellulose. Biofuels Bioproducts and Biorefining, 2012, 6 (5): 561-579.

[115] Zhao X B, Zhang L H, Liu D H. Biomass recalcitrance. Part I: the chemical compositions and physical structures affecting the enzymatic hydrolysis of lignocellulose. Biofuels Bioproducts and Biorefining, 2012, 6 (4): 465-482.

[116] Heeres H J, Girisuta B, Janssen L P B M. A kinetic study on the conversion of glucose to levulinic acid. Chemical Engineering Research and Design, 2006, 84 (A5): 339-349.

[117] Heeres H J, Rasrendra C B, Makertihartha I G B N, Adisasmito S. Green chemicals from D-glucose: systematic studies on catalytic effects of inorganic salts on the chemo-selectivity and yield in aqueous solutions. Topics in Catalysis, 2010, 53 (15-18): 1241-1247.

[118] Pilath H M, Nimlos M R, Mittal A, Himmel M E, Johnson D K. Glucose Reversion Reaction Kinetics. Journal of Agricultural and Food Chemistry, 2010, 58 (10): 6131-6140.

[119] Dee S J, Bell A T. A study of the acid-catalyzed hydrolysis of cellulose dissolved in ionic liquids and the factors influencing the dehydration of glucose and the formation of humins. ChemSusChem, 2011, 4 (8): 1166-1173.

[120] Ehara K, Saka S. Decomposition behavior of cellulose in supercritical water, subcritical water, and their combined treatments. Journal of Wood Science, 2005, 51 (2): 148-153.

[121] Daorattanachai P, Khemthong P, Viriya-Empikul N, Laosiripojana N, Faungnawakij K. Conversion of fructose, glucose, and cellulose to 5-hydroxymethylfurfural by alkaline earth phosphate catalysts in hot compressed water. Carbohydrate Research, 2012, 363 58-61.

[122] Swatloski R P, Spear S K, Holbrey J D, Rogers R D. Dissolution of cellose with ionic liquids. Journal of the American Chemical Society, 2002, 124 (18): 4974-4975.

[123] Zhu S D, Wu Y X, Chen Q M, Yu Z N, Wang C W, Jin S W, Ding Y G, Wu G. Dissolution of cellulose with ionic liquids and its application: a mini-review. Green Chemistry, 2006, 8 (4): 325-327.

[124] Zavrel M, Bross D, Funke M, Buchs J, Spiess A C. High-throughput screening for ionic liquids dissolving (ligno-) cellulose. Bioresource Technology, 2009, 100 (9): 2580-2587.

[125] Hsu W H, Lee Y Y, Peng W H, Wu K C W. Cellulosic conversion in ionic liquids (ILs): Effects of H_2O/cellulose molar ratios, temperatures, times, and different ILs on the production of monosaccharides and 5-hydroxymethylfurfural (HMF). Catalysis Today, 2011, 174 (1): 65-69.

[126] Yu S, Brown H M, Huang X W, Zhou X D, Amonette J E, Zhang Z C. Single-step conversion of cellulose to 5-hydroxymethylfurfural (HMF), a versatile platform chemical. Applied Catalysis A: General, 2009, 361 (1-2): 117-122.

[127] Li C Z, Zhang Z H, Zhao Z B K. Direct conversion of glucose and cellulose to 5-hydroxymethylfurfural in ionic liquid under microwave irradiation. Tetrahedron Letters, 2009, 50 (38): 5403-5405.

[128] Ding Z D, Shi J C, Xiao J J, Gu W X, Zheng C G, Wang H J. Catalytic conversion of cellulose

to 5-hydroxymethyl furfural using acidic ionic liquids and co-catalyst. Carbohydrate Polymers, 2012, 90 (2): 792-798.

[129] Zhang Z C, Yu S, Brown H M, Huang X W, Zhou X D, Amonette J E. Single-step conversion of cellulose to 5-hydroxymethylfurfural (HMF), a versatile platform chemical. Applied Catalysis A: General, 2009, 361 (1-2): 117-122.

[130] Zhang Z, Du B, Zhang L J, Da Y X, Quan Z J, Yang L J, Wang X C. Conversion of carbohydrates into 5-hydroxymethylfurfural using polymer bound sulfonic acids as efficient and recyclable catalysts. Rsc Advances, 2013, 3 (24): 9201-9205.

[131] Tian G, Tong X L, Cheng Y, Xue S. Tin-catalyzed efficient conversion of carbohydrates for the production of 5-hydroxymethylfurfural in the presence of quaternary ammonium salts. Carbohydrate Research, 2013, 370 33-37.

[132] Shen F, Sun S, Zhang X, Yang J, Qiu M, Qi X. Mechanochemical-assisted production of 5-hydroxymethylfurfural from high concentration of cellulose. Cellulose, 2020, 27 (6): 3013-3023.

[133] Zhang Z H, Wang Q A, Xie H B, Liu W J, Zhao Z B. Catalytic conversion of carbohydrates into 5-hydroxymethylfurfural by germanium (IV) Chloride in ionic liquids. ChemSusChem, 2011, 4 (1): 131-138.

[134] Tan M X, Zhao L, Zhang Y G. Production of 5-hydroxymethyl furfural from cellulose in $CrCl_2$/Zeolite/BMIMCl system. Biomass and Bioenergy, 2011, 35 (3): 1367-1370.

[135] Qi X H, Watanabe M, Aida T M, Smith R L. Fast transformation of glucose and di-/polysaccharides into 5-hydroxymethylfurfural by microwave heating in an ionic liquid/catalyst system. ChemSusChem, 2010, 3 (9): 1071-1077.

[136] Mcneff C V, Nowlan D T, Mcneff L C, Yan B W, Fedie R L. Continuous production of 5-hydroxymethylfurfural from simple and complex carbohydrates. Applied Catalysis A: General, 2010, 384 (1-2): 65-69.

[137] Bozell J J. Chemistry connecting biomass and petroleum processing with a chemical bridge. Science, 2010, 329 (5991): 522-523.

[138] Bozell J J, Moens L, Elliott D C, Wang Y, Neuenscwander G G, Fitzpatrick S W, Bilski R J, Jarnefeld J L. Production of levulinic acid and use as a platform chemical for derived products. Resources Conservation and Recycling, 2000, 28 (3-4): 227-239.

[139] Mascal M, Dutta S, Gandarias I. Hydrodeoxygenation of the angelica lactone dimer, a cellulose-based feedstock: simple, high-yield synthesis of branched $C_7 \sim C_{10}$ gasoline-like hydrocarbons. Angewandte Chemie-International Edition, 2014, 53 (7): 1854-1857.

[140] Xin J Y, Zhang S J, Yan D, Ayodele O, Lu X M, Wang J. Formation of C—C bonds for the bio-alkanes production at mild conditions. Green Chemistry, 2014.

[141] Rackemann D W, Doherty W O S. The conversion of lignocellulosics to levulinic acid. Biofuels, Bioproducts and Biorefining, 2011, 5 (2): 198-214.

[142] Isikgor F H, Becer C R. Lignocellulosic biomass: a sustainable platform for the production of bio-based chemicals and polymers. Polymer Chemistry, 2015, 6 (25): 4497-4559.

[143] Pileidis F D, Titirici M M. Levulinic acid biorefineries: new challenges for efficient utilization of biomass. ChemSusChem, 2016, 9 (6): 562-582.

[144] Horvat J, Klaic B, Metelko B, Sunjic V. Mechanism of levulinic acid formation. Tetrahedron Letters, 1985, 26 (17): 2111-2114.

[145] Zhang J, Wu S B, Li B, Zhang H D. Advances in the catalytic production of valuable levulinic acid derivatives. ChemCatChem, 2012, 4 (9): 1230-1237.

[146] Chang C, Ma X, Cen P. Kinetics of levulinic acid formation from glucose decomposition at high temperature. Chinese Journal of Chemical Engineering, 2006, 14 (5): 708-712.

[147] Davidek T, Robert F, Devaud S, Vera F A, Blank I. Sugar fragmentation in the maillard reaction cascade: formation of short-chain carboxylic acids by a new oxidative α-dicarbonyl cleavage pathway. Journal of Agricultural and Food Chemistry, 2006, 54 (18): 6677-6684.

[148] Davidek T, Devaud S, Robert F, Blank I. Sugar fragmentation in the maillard reaction cascade: isotope labeling studies on the formation of acetic acid by a hydrolytic β-dicarbonyl cleavage mechanism. Journal of Agricultural and Food Chemistry, 2006, 54 (18): 6667-6676.

[149] Gobert J, Glomb M A. Degradation of glucose: reinvestigation of reactive α-dicarbonyl compounds. Journal of Agricultural and Food Chemistry, 2009, 57 (18): 8591-8597.

[150] Tarabanko V E, Aralova M Y C V, Kuznetsov B N. Kinetics of levulinic acid formation from carbohydrates at moderate temperatures. Reaction Kinetics and Catalysis Letters, 2002, 75 (1): 117-126.

[151] Takeuchi Y, Jin F M, Tohji K, Enomoto H. Acid catalytic hydrothermal conversion of carbohydrate biomass into useful substances. Journal of Materials Science. 2008, 43 (7): 2472-2475.

[152] Fachri B A, Abdilla R M, Bovenkamp H H V D, Rasrendra C B, Heeres H J. Experimental and kinetic modeling studies on the sulfuric acid catalyzed conversion of D-Fructose to 5-hydroxymethylfurfural and levulinic acid in water. Acs Sustainable Chemistry and Engineering. 2015, 3 (12): 3024-3034.

[153] Heeres H J, Girisuta B, Janssen L P B M. Kinetic study on the acid-catalyzed hydrolysis of cellulose to levulinic acid. Industrial and Engineering Chemistry Research, 2007, 46 (6): 1696-1708.

[154] Morone A, Apte M, Pandey R A. Levulinic acid production from renewable waste resources: Bottlenecks, potential remedies, advancements and applications. Renewable and Sustainable Energy Reviews, 2015, 51: 548-565.

[155] Kang S M, Fu J X, Zhang G. From lignocellulosic biomass to levulinic acid: A review on acid-catalyzed hydrolysis. Renewable and Sustainable Energy Reviews, 2018, 94: 340-362.

[156] Cao X, Peng X, Sun S, Zhong L, Chen W, Wang S, Sun R C. Hydrothermal conversion of xylose, glucose, and cellulose under the catalysis of transition metal sulfates. Carbohydrate Polymers, 2015, 118: 44-51.

[157] Zheng X, Zhi Z, Gu X, Li X, Zhang R, Lu X. Kinetic study of levulinic acid production from corn stalk at mild temperature using $FeCl_3$ as catalyst. Fuel, 2017, 187: 261-267.

[158] Yang F, Fu J, Mo J, Lu X Y. Synergy of Lewis and Brønsted acids on catalytic hydrothermal decomposition of hexose to levulinic acid. Energy and Fuels, 2013, 27 (11): 6973-6978.

[159] Choudhary V, Mushrif S H, Ho C, Anderko A, Nikolakis V, Marinkovic N S, Frenkel A I, Sandler S I, Vlachos D G. Insights into the interplay of Lewis and Brønsted acid catalysts in glucose and fructose conversion to 5-(hydroxymethyl) furfural and levulinic acid in aqueous media. Journal of the American Chemical Society, 2013, 135 (10): 3997-4006.

[160] Ramli N A S, Amin N A S. Kinetic study of glucose conversion to levulinic acid over Fe/HY zeolite catalyst. Chemical Engineering Journal, 2016, 283: 150-159.

[161] Liu Y, Li H, He J, Zhao W, Yang T, Yang S. Catalytic conversion of carbohydrates to levulinic acid with mesoporous niobium-containing oxides. Catalysis Communications, 2017, 93: 20-24.

[162] Joshi S S, Zodge A D, Pandare K V, Kulkarni B D. Efficient conversion of cellulose to levulinic acid by hydrothermal treatment using zirconium dioxide as a recyclable solid acid catalyst. Industrial and Engineering Chemistry Research, 2014, 53: 18796-18805.

[163] Wang P, Zhan S H, Yu H B, Xue X F, Hong N. The effects of temperature and catalysts on the pyrolysis of industrial wastes (herb residue). Bioresource Technology, 2010, 101 (9): 3236-3241.

[164] Weingarten R, Kim Y T, Tompsett G A, Fernandez A, Han K S, Hagaman E W, Conner W C, Dumesic J A, Huber G W. Conversion of glucose into levulinic acid with solid metal (IV) phosphate catalysts. Journal of Catalysis, 2013, 304: 123-134.

[165] Zuo Y, Zhang Y, Fu Y. Catalytic conversion of cellulose into levulinic acid by a sulfonated chloromethyl polystyrene solid acid catalyst. ChemCatChem, 2014, 6 (3): 753-757.

[166] Alonso D M, Gallo J M R, Mellmer M A, Wettstein S G, Dumesic J A. Direct conversion of cellulose to levulinic acid and gamma-valerolactone using solid acid catalysts. Catalysis Science and Technology, 2013, 3 (4): 927-931.

[167] Weingarten R, Conner W C, Huber G W. Production of levulinic acid from cellulose by hydrothermal decomposition combined with aqueous phase dehydration with a solid acid catalyst. Energy and Environmental Science, 2012, 5 (6): 7559-7574.

[168] Van De Vyver S, Thomas J, Geboers J, Keyzer S, Smet M, Dehaen W, Jacobs P A, Sels B F. Catalytic production of levulinic acid from cellulose and other biomass-derived carbohydrates with sulfonated hyperbranched poly (arylene oxindole) s. Energy and Environmental Science, 2011, 4 (9): 3601.

[169] Xu Y, Liu G, Fu J, Kang S, Xiao Y, Yang P, Liao W. Catalytic hydrolysis of cellulose to levulinic acid by partly replacing sulfuric acid with Nafion® NR50 catalyst. Biomass Conversion and Biorefinery, 2019, 9 (3): 609-616.

[170] Hegner J, Pereira K C, Deboef B, Lucht B L. Conversion of cellulose to glucose and levulinic acid via solid-supported acid catalysis. Tetrahedron Letters, 2010, 51 (17): 2356-2358.

[171] Wang K, Jiang J, Liang X, Wu H, Xu J. Direct conversion of cellulose to levulinic acid over multifunctional sulfonated humins in sulfolane-water solution. Acs Sustainable Chemistry and Engineering, 2018, 6 (11): 15092-15099.

[172] Lin H F, Strull J, Liu Y, Karmiol Z, Plank K, Miller G, Guo Z H, Yang L S. High yield production of levulinic acid by catalytic partial oxidation of cellulose in aqueous media. Energy and

Environmental Science, 2012, 5 (12): 9773-9777.

[173] Lin H, Strull J, Liu Y, Karmiol Z, Plank K, Miller G, Guoc Z, Yang L. High yield production of levulinic acid by catalytic partial oxidation of cellulose in aqueous media. Energy and Environmental Science, 2012, 5: 9773-9777.

[174] Kang S, Yu J. An intensified reaction technology for high levulinic acid concentration from lignocellulosic biomass. Biomass and Bioenergy, 2016, 95: 214-220.

[175] Joshi H, Moser B R, Toler J, Smith W F, Walker T. Ethyl levulinate: a potential bio-based diluent for biodiesel which improves cold flow properties. Biomass and Bioenergy, 2011, 35 (7): 3262-3266.

[176] Christensen E, Williams A, Paul S, Burton S, Mccormick R L. Properties and performance of levulinate esters as diesel blend components. Energy and Fuels, 2011, 25 (11): 5422-5428.

[177] Lei T, Wang Z, Chang X, Lin L, Yan X, Sun Y, Shi X, He X, Zhu J. Performance and emission characteristics of a diesel engine running on optimized ethyl levulinate-biodiesel-diesel blends. Energy, 2016, 95: 29-40.

[178] Climent M J, Corma A, Iborra S. Conversion of biomass platform molecules into fuel additives and liquid hydrocarbon fuels. Green Chemistry, 2014, 16 (2): 516-547.

[179] 常春, 邓琳, 戚小各, 白净, 方书起. 固体催化剂在生物质合成乙酰丙酸和乙酰丙酸酯中的应用研究进展. 林产化学与工业, 2017, 37 (2): 11-21.

[180] Peng L C, Lin L, Li H. Conversion of biomass into levulinate esters as novel energy chemicals. Progress in Chemistry, 2012, 24 (5): 801-809.

[181] Nandiwale K Y, Sonar S K, Niphadkar P S, Joshi P N, Deshpande S S, Patil V S, Bokade V V. Catalytic upgrading of renewable levulinic acid to ethyl levulinate biodiesel using dodecatungstophosphoric acid supported on desilicated H-ZSM-5 as catalyst. Applied Catalysis A: General, 2013, 460-461: 90-98.

[182] Gómez Bernal H, Benito P, Rodríguez-Castellón E, Raspolli Galletti A M, Funaioli T. Synthesis of isopropyl levulinate from furfural: Insights on a cascade production perspective. Applied Catalysis A: General, 2019, 575: 111-119.

[183] Démolis A, Essayem N, Rataboul F. Synthesis and applications of alkyl levulinates. Acs Sustainable Chemistry and Engineering, 2014, 2 (6): 1338-1352.

[184] Huang Y B, Yang T, Lin Y T, Zhu Y Z, Li L C, Pan H. Facile and high-yield synthesis of methyl levulinate from cellulose. Green Chemistry, 2018, (20): 1323-1334

[185] Dusselier M, Wouwe P V, Dewaele A, Makshina E, Sels B F. Lactic acid as a platform chemical in the biobased economy: the role of chemocatalysis. Energy and Environmental Science, 2013, 6: 1415-1442

[186] Maki-Arvela P, Simakova I L, Salmi T, Murzin D Y. Production of lactic acid/lactates from biomass and their catalytic transformations to commodities. Chemical Reviews, 2014, 114 (3): 1909-1971.

[187] Fan Y X, Zhou C H, Zhu X H. Selective catalysis of lactic acid to produce commodity chemicals. Catalysis Reviews: Science and Engineering, 2009, 51 (3): 293-324.

[188] Madhavan Nampoothiri K, Nair N R, John R P. An overview of the recent developments in polylactide (PLA) research. Bioresource Technology, 2010, 101 (22): 8493-8501.

[189] Evans W L. Some less familiar aspects of carbohydrate chemistry. Chemical Reviews, 1942, 537-560.

[190] Srokol Z, Bouche A G, Van Estrik A, Strik R C J, Maschmeyer T, Peters J A. Hydrothermal upgrading of biomass to biofuel: studies on some monosaccharide model compounds. Carbohydrate Research, 2004, 339 (10): 1717-1726.

[191] Holm M S, Saravanamurugan S, Taarning E. Conversion of sugars to lactic acid derivatives using heterogeneous zeotype catalysts. Science, 2010, 328 (5978): 602-605.

[192] Jin F M, Zhou Z Y, Enomoto H, Moriya T, Higashijima H. Conversion mechanism of cellulosic biomass to lactic acid in subcritical water and acid-base catalytic effect of subcritical water. Chemistry Letters, 2004, 33 (2): 126-127.

[193] Deng W, Zhang Q, Wang Y. Catalytic transformation of cellulose and its derived carbohydrates into chemicals involving C—C bond cleavage. Journal of Energy Chemistry, 2015, 24 (5): 595-607.

[194] Krochta J M, Tillin S J, Hudson J S. Degradation of polysaccharides in alkaline-solution to organic-acids-product characterization and identification. Journal of Applied Polymer Science, 1987, 33 (4): 1413-1425.

[195] Kabyemela B M, Adschiri T, Malaluan R M, Arai K. Glucose and fructose decomposition in subcritical and supercritical water: detailed reaction pathway, mechanisms, and kinetics. Industrial and Engineering Chemistry Research, 1999, 38 (8): 2888-2895.

[196] Gibbs M. On the mechanism of the chemical formation of lactic acid from glucose studied with C_{14} labeled glucose. Journal of the American Chemical Society, 1950, 72: 3964-3965.

[197] Whistler R L, Bemiller J N. Alkaline degradation of polysaccharides. Advances in Carbohydrate Chemistry, 1958, 13: 289-329.

[198] Tolborg S, Meier S, Sadaba I, Elliot S G, Kristensen S K, Saravanamurugan S, Riisager A, Fristrup P, Skrydstrup T, Taarning E. Tin-containing silicates: identification of a glycolytic riathway via 3-deoxyglucosone. Green Chemistry, 2016, 18 (11): 3360-3369.

[199] Wang A Q, Zhang T. One-pot conversion of cellulose to ethylene glycol with multifunctional tungsten-based catalysts. Accounts of Chemical Research, 2013, 46 (7): 1377-1386.

[200] Van Zandvoort I, Wang Y H, Rasrendra C B, Van Eck E R H, Bruijnincx P C A, Heeres H J, Weckhuysen B M. Formation, molecular structure, and morphology of humins in biomass conversion: influence of feedstock and processing conditions. ChemSusChem, 2013, 6 (9): 1745-1758.

[201] Knill C J, Kennedy J F. Degradation of cellulose under alkaline conditions. Carbohydrate Polymers, 2003, 51 (3): 281-300.

[202] Yan X Y, Jin F M, Tohji K, Kishita A, Enomoto H. Hydrothermal conversion of carbohydrate biomass to lactic acid. Aiche Journal, 2010, 56 (10): 2727-2733.

[203] Esposito D, Antonietti M. Chemical conversion of sugars to lactic acid by alkaline hydrothermal processes. ChemSusChem, 2013, 6 (6): 989-992.

[204] Zhang S P, Jin F M, Hu J J, Huo Z B. Improvement of lactic acid production from cellulose

with the addition of Zn/Ni/C under alkaline hydrothermal conditions. Bioresource Technology, 2011, 102 (2): 1998-2003.

[205] Wang F W, Huo Z B, Wang Y Q, Jin F M. Hydrothermal conversion of cellulose into lactic acid with nickel catalyst. Research on Chemical Intermediates, 2011, 37 (2-5): 487-492.

[206] Li L, Shen F, Smith R L, Qi X. Quantitative chemocatalytic production of lactic acid from glucose under anaerobic conditions at room temperature. Green Chemistry, 2017, 19 (1): 76-81.

[207] Liu Z, Li W, Pan C Y, Chen P, Lou H, Zheng X M. Conversion of biomass-derived carbohydrates to methyl lactate using solid base catalysts. Catalysis Communications, 2011, 15 (1): 82-87.

[208] Onda A, Ochi T, Kajiyoshi K, Yanagisawa K. Lactic acid production from glucose over activated hydrotalcites as solid base catalysts in water. Catalysis Communications, 2008, 9 (6): 1050-1053.

[209] Sánchez C, Egüés I, García A, Llano-Ponte R, Labidi J. Lactic acid production by alkaline hydrothermal treatment of corn cobs. Chemical Engineering Journal. 2012, 181-182: 655-660.

[210] He T, Jiang Z, Wu P, Yi J, Li J, Hu C. Fractionation for further conversion: from raw corn stover to lactic acid. Scientific Reports, 2016, 6 (1): 38623.

[211] Duo J, Zhang Z S, Yao G D, Huo Z B, Jin F M. Hydrothermal conversion of glucose into lactic acid with sodium silicate as a base catalyst. Catalysis Today, 2016, 263: 112-116.

[212] Onda A, Ochi T, Kajiyoshi K, Yanagisawa K. A new chemical process for catalytic conversion Of D-glucose into lactic acid and gluconic acid. Applied Catalysis A: General, 2008, 343 (1-2): 49-54.

[213] Hayashi Y, Sasaki Y. Tin-catalyzed conversion of trioses to alkyl lactates in alcohol solution. Chemical Communications, 2005, (21): 2716-2718.

[214] Bicker M, Endres S, Ott L, Vogel H. Catalytical conversion of carbohydrates in subcritical water: A new chemical process for lactic acid production. Journal of Molecular Catalysis A: Chemical, 2005, 239 (1-2): 151-157.

[215] Wang Y L, Deng W P, Wang B J, Zhang Q H, Wan X Y, Tang Z C, Wang Y, Zhu C, Cao Z X, Wang G C, Wan H L. Chemical synthesis of lactic acid from cellulose catalysed by lead (II) ions in water. Nature Communications, 2013, 4: 1-7.

[216] Lei X, Wang F F, Liu C L, Yang R Z, Dong W S. One-pot catalytic conversion of carbohydrate biomass to lactic acid using an $ErCl_3$ catalyst. Applied Catalysis A: General, 2014, 482: 78-83.

[217] Wang F F, Liu C L, Dong W S. Highly efficient production of lactic acid from cellulose using lanthanide triflate catalysts. Green Chemistry, 2013, 15 (8): 2091-2095.

[218] Tang Z C, Deng W P, Wang Y L, Zhu E Z, Wan X Y, Zhang Q H, Wang Y. Transformation of cellulose and its derived carbohydrates into formic and lactic acids catalyzed by vanadyl cations. ChemSusChem, 2014, 7 (6): 1557-1567.

[219] Deng W, Wang P, Wang B, Wang Y, Yan L, Li Y, Zhang Q, Cao Z, Wang Y. Transformation of cellulose and related carbohydrates into lactic acid with bifunctional Al (III)-Sn (II) catalysts. Green Chemistry, 2018, 20 735-744

[220] Zhou L P, Wu L, Li H J, Yang X M, Su Y L, Lu T L, Xu J. A facile and efficient method to improve the selectivity of methyl lactate in the chemocatalytic conversion of glucose catalyzed by homogeneous Lewis acid. Journal of Molecular Catalysis A: Chemical, 2014, 388: 74-80.

[221] Lv F H, Bi R, Liu Y H, Li W S, Zhou X P. The synthesis of methyl lactate and other methyl oxygenates from cellulose. Catalysis Communications, 2014, 49: 78-81.

[222] Nemoto K, Hirano Y, Hirata K, Takahashi T, Tsuneki H, Tominaga K, Sato K. Cooperative In-Sn catalyst system for efficient methyl lactate synthesis from biomass-derived sugars. Applied Catalysis B: Environmental, 2016, 183: 8-17.

[223] Lu X, Fu J, Langrish T, Lu X. Simultaneous catalytic conversion of C_6 and C_5 sugars to methyl lactate in near-critical methanol with metal chlorides. Bioresources, 2018, 13 (2): 3627-3641.

[224] Liu D, Kim K H, Sun J, Simmons B A, Singh S. Cascade production of lactic acid from universal types of sugars catalyzed by lanthanum triflate. ChemSusChem, 2018, 11 (3): 598-604.

[225] 岳孝阳, 李林峰, 李吉凡, 杨荣榛, 董文生. 有机碱协同 $SnCl_2 \cdot 2H_2O$ 催化葡萄糖转化制备乳酸甲酯. 工业催化, 2017, 25 (11): 65-68.

[226] Bayu A, Yoshida A, Karnjanakom S, Kusakabe K, Hao X, Prakoso T, Abudula A, Guan G. Catalytic conversion of biomass derivatives to lactic acid with increased selectivity in an aqueous tin (ii) chloride. Green Chemistry, 2018, 20: 4112-4119

[227] Wang J, Yao G, Jin F. One-pot catalytic conversion of carbohydrates into alkyl lactates with Lewis acids in alcohols. Molecular Catalysis, 2017, 435 82-90.

[228] Janssen K P F, Paul J S, Sels B F, Jacobsa P A. Glyoxylase biomimics zeolite catalyzed conversion of trioses. Studies in Surface Science and Catalysis, 2007, 170: 1222-1227.

[229] Taarning E, Saravanamurugan S, Holm M S, Xiong J M, West R M, Christensen C H. Zeolite-catalyzed isomerization of triose sugars. ChemSusChem, 2009, 2 (7): 625-627.

[230] Zhang J, Wang L, Wang G, Chen F, Zhu J, Wang C, Bian C, Pan S, Xiao F S. Hierarchical Sn-Beta Zeolite catalyst for the conversion of sugars to alkyl lactates. Acs Sustainable Chemistry and Engineering, 2017, 5 (4): 3123-3131.

[231] Yang X, Bian J, Huang J, Xin W, Lu T, Chen C, Su Y, Zhou L, Wang F, Xu J. Fluoride-free and low concentration template synthesis of hierarchical Sn-Beta zeolites: efficient catalysts for conversion of glucose to alkyl lactate. Green Chemistry, 2017, 19 (3): 692-701.

[232] Yang X, Liu Y, Li X, Ren J, Zhou L, Lu T, Su Y. Synthesis of Sn-containing nanosized Beta zeolite as efficient catalyst for transformation of glucose to methyl lactate. Acs Sustainable Chemistry and Engineering, 2018, 6 (7): 8256-8265.

[233] Wang J C, Masui Y, Onaka M. Conversion of triose sugars with alcohols to alkyl lactates catalyzed by Brønsted acid tin ion-exchanged montmorillonite. Applied Catalysis B: Environmental. 2011, 107 (1-2): 135-139.

[234] Murillo B, Sánchez A, Sebastián V, Casado C, Iglesia O D L, López-Ram-De-Viu M P, Téllez C, Coronas J. Conversion of glucose to lactic acid derivatives with mesoporous Sn-MCM-41 and microporous titanosilicates. Journal of Chemical Technology and Biotechnology, 2014, 89 (9): 1344-1350.

[235] Liu Z, Feng G, Pan C, Li W, Chen P, Lou H, Zheng X. Conversion of Biomass-Derived Carbohydrates to Methyl Lactate Using Sn-MCM-41 and SnO_2/SiO_2. Chinese Journal of Catalysis, 2012, 33 (10): 1696-1705.

[236] Yang X M, Wu L, Wang Z, Bian J J, Lu T L, Zhou L P, Chen C, Xu J. Conversion of dihydroxyacetone to methyl lactate catalyzed by highly active hierarchical Sn-USY at room temperature. Catalysis Science and Technology, 2016, 6 (6): 1757-1763.

[237] Guo Q, Fan F, Pidko E A, Van Der Graaff W N, Feng Z, Li C, Hensen E J. Highly active and recyclable Sn-MWW zeolite catalyst for sugar conversion to methyl lactate and lactic acid. ChemSusChem, 2013, 6 (8): 1352-1356.

[238] Pang J, Zheng M, Li X, Song L, Sun R, Sebastian J, Wang A, Wang J, Wang X, Zhang T. Catalytic Conversion of Carbohydrates to Methyl Lactate Using Isolated Tin Sites in SBA-15. Chemistryselect, 2017, 2 (1): 309-314.

[239] Godard N, Vivian A, Fusaro L, Cannavicci L, Aprile C, Debecker D P. High-yield synthesis of ethyl lactate with mesoporous tin silicate catalysts prepared by an aerosol-assisted sol-gel process. ChemCatChem, 2017, 9 (12): 2211-2218.

[240] De Clippel F, Dusselier M, Van Rompaey R, Vanelderen P, Dijkmans J, Makshina E, Giebeler L, Oswald S, Baron G V, Denayer J F M, Pescarmona P P, Jacobs P A, Sels B F. Fast and selective sugar conversion to alkyl lactate and lactic acid with bifunctional carbon-silica catalysts. Journal of the American Chemical Society, 2012, 134 (24): 10089-10101.

[241] Tolborg S, Sadaba I, Osmundsen C M, Fristrup P, Holm M S, Taarning E. Tin-containing silicates: alkali salts improve methyl lactate yield from sugars. ChemSusChem, 2015, 8 (4): 613-617.

[242] Dong W, Shen Z, Peng B, Gu M, Zhou X, Xiang B, Zhang Y. Selective chemical conversion of sugars in aqueous solutions without alkali to lactic acid over a Zn-Sn-Beta Lewis acid-base catalyst. Scientific Reports, 2016, 6 1-8.

[243] Xia M, Dong W, Gu M, Chang C, Shen Z, Zhang Y. Synergetic effects of bimetals in modified beta zeolite for lactic acid synthesis from biomass-derived carbohydrates. Rsc Advances, 2018, 8 (16): 8965-8975.

[244] Kong L, Shen Z, Zhang W, Xia M, Gu M, Zhou X, Zhang Y. Conversion of sucrose into lactic acid over functionalized Sn-Beta zeolite catalyst by 3-aminopropyltrimethoxysilane. ACS Omega, 2018, 3 (12): 17430-17438.

[245] Elliot S G, Tolborg S, Madsen R, Taarning E, Meier S. Effects of alkali-metal ions and counter ions in Sn-Beta-catalyzed carbohydrate conversion. ChemSusChem, 2018, 11 (7): 1198-1203.

[246] Orazov M, Davis M E. Tandem catalysis for the production of alkyl lactates from ketohexoses at moderate temperatures. Proceedings of the National Academy of Sciences of the United States of America, 2015, 112 (38): 11777-11782.

[247] Lau K S, Chia C H, Chin S X, Chook S W, Zakaria S, Juan J C. Conversion of glucose into lactic acid using silica-supported zinc oxide as solid acid catalyst. Pure and Applied Chemistry, 2018, 90 (6): 1035-1043.

[248] Cao D, Cai W, Tao W, Zhang S, Wang D, Huang D. Lactic acid production from glucose over a novel Nb_2O_5 nanorod catalyst. Catalysis Letters, 2017, 147 (4): 926-933.

[249] Shi N, Liu Q Y, He X, Cen H, Ju R M, Zhang Y L, Ma L L. Production of lactic acid from cellulose catalyzed by easily prepared solid $Al_2(WO_4)_3$. Bioresource Technology Reports, 2019, 5: 66-73.

[250] Feliczak-Guzik A, Sprynskyy M, Nowak I, Jaroniec M, Buszewski B. Application of novel hierarchical niobium-containing zeolites for synthesis of alkyl lactate and lactic acid. Journal of Colloid and Interface Science, 2018, 516: 379-383.

[251] Wang X, Liang F, Huang C, Li Y, Chen B. Highly active tin (IV) phosphate phase transfer catalysts for the production of lactic acid from triose sugars. Catalysis Science and Technology. 2015, 5 (9): 4410-4421.

[252] Takagaki A, Goto H, Kikuchi R, Oyama S T. Silica-supported chromia-titania catalysts for selective formation of lactic acid from a triose in water. Applied Catalysis A: General, 2019, 570: 200-208.

[253] Hou G H, Yan L F. Synthesis of $Pb(OH)_2$/rGO catalyst for conversion of sugar to lactic acid in water. Chinese Journal of Chemical Physics, 2015, 28 (4): 533-538.

[254] Yang L S, Su J, Carl S, Lynam J G, Yang X K, Lin H F. Catalytic conversion of hemicellulosic biomass to lactic acid in pH neutral aqueous phase media. Applied Catalysis B: Environmental, 2015, 162: 149-157.

[255] Yang L, Yang X, Tian E, Lin H. Direct Conversion of Cellulose into Ethyl Lactate in Supercritical Ethanol—Water Solutions. ChemSusChem, 2016, 9 (1): 36-41.

[256] Yang L S, Yang X K, Tian E, Vattipalli V, Fan W, Lin H F. Mechanistic insights into the production of methyl lactate by catalytic conversion of carbohydrates on mesoporous Zr-SBA-15. Journal of Catalysis, 2016, 333: 207-216.

[257] Verma D, Insyani R, Suh Y W, Kim S M, Kim S K, Kim J. Direct conversion of cellulose to high-yield methyl lactate over Ga-doped Zn/H-nanozeolite Y catalysts in supercritical methanol. Green Chemistry, 2017, 19 (8): 1969-1982.

[258] Wang F F, Wu H Z, Ren H F, Liu C L, Xu C L, Dong W S. Er/beta-zeolite-catalyzed one-pot conversion of cellulose to lactic acid. Journal of Porous Materials, 2017, 24 (3): 697-706.

[259] Wang F F, Liu J, Li H, Liu C L, Yang R Z, Dong W S. Conversion of cellulose to lactic acid catalyzed by erbium-exchanged montmorillonite K10. Green Chemistry, 2015, 17 (4): 2455-2463.

[260] Li H, Ren H F, Zhao B W, Liu C L, Yang R Z, Dong W S. Production of lactic acid from cellulose catalyzed by alumina-supported Er_2O_3 catalysts. Research on Chemical Intermediates, 2016, 42 (9): 7199-7211.

[261] Zhao B, Yue X, Li H, Li J, Liu C L, Xu C, Dong W S. Lanthanum-modified phosphomolybdic acid as an efficient catalyst for the conversion of fructose to lactic acid. Reaction Kinetics, Mechanisms and Catalysis, 2018, 125 (1): 55-69.

[262] Wattanapaphawong P, Reubroycharoen P, Yamaguchi A. Conversion of cellulose into lactic acid using zirconium oxide catalysts. Rsc Advances, 2017, 7 (30): 18561-18568.

[263] Wattanapaphawong P, Sato O, Sato K, Mimura N, Reubroycharoen P, Yamaguchi A. Conversion of cellulose to lactic acid by using ZrO_2-Al_2O_3 catalysts. Catalysts, 2017, 7 (7): 221-231.

[264] Yamaguchi S, Yabushita M, Kim M, Hirayama J, Motokura K, Fukuoka A, Nakajima K. Catalytic conversion of biomass-derived carbohydrates to methyl lactate by acid-base bifunctional γ-Al_2O_3. Acs Sustainable Chemistry and Engineering, 2018, 6 (7): 8113-8117.

[265] Lyu X, Xu L, Wang J, Lu X. New insights into the NiO catalytic mechanism on the conversion of fructose to methyl lactate. Catalysis Communications, 2019, 119: 46-50.

[266] Dusselier M, Sels B F. Selective catalysis for cellulose conversion to lactic acid and other α-hydroxy acids. Selective Catalysis for Renewable Feedstocks and Chemicals, 2014, 353: 85-125.

[267] Sølvhøj A, Taarning E, Madsen R. Methyl vinyl glycolate as a diverse platform molecule. Green Chemistry, 2016, 18 (20): 5448-5455.

[268] Dusselier M, Van Wouwe P, De Clippel F, Dijkmans J, Gammon D W, Sels B F. Mechanistic insight into the conversion of tetrose sugars to novel α-hydroxy acid platform molecules. ChemCatChem, 2013, 5 (2): 569-575.

[269] Elliot S G, Taarning E, Madsen R, Meier S. NMR Isotope Tracking Reveals Cascade Steps in Carbohydrate Conversion by Sn-Beta. ChemCatChem, 2018, 10 (6): 1414-1419.

[270] Chen H S, Wang A, Sorek H, Lewis J D, Roman-Leshkov Y, Bell A T. Production of hydroxyl-rich acids from xylose and glucose using Sn-BEA Zeolite. Chemistryselect, 2016, 1 (14): 4167-4172.

[271] Elliot S G, Andersen C, Tolborg S, Meier S, Sádaba I, Daugaard A E, Taarning E. Synthesis of a novel polyester building block from pentoses by tin-containing silicates. Rsc Advances, 2017, 7 (2): 985-996.

[272] Holm M S, Pagán-Torres Y J, Saravanamurugan S, Riisager A, Dumesic J A, Taarning E. Sn-Beta catalysed conversion of hemicellulosic sugars. Green Chemistry, 2012, 14 702-706

[273] Clercq R D, Dusselier M, Christiaens C, Dijkmans J, Iacobescu R I, Pontikes Y, Sels B F. Confinement effects in Lewis acid-catalyzed sugar conversion: steering toward functional polyester building blocks. Acs Catalysis, 2015, 5 (10): 5803-5811.

[274] Shrotri A, Kobayashi H, Fukuoka A. Cellulose depolymerization over heterogeneous catalysts. Accounts of Chemical Research, 2018, 51 (3): 761-768.

[275] Zhang J, Li J B, Wu S B, Liu Y. Advances in the catalytic production and utilization of sorbitol. Industrial and Engineering Chemistry Research, 2013, 52 (34): 11799-11815.

[276] Luo C, Wang S, Liu H. Cellulose conversion into polyols catalyzed by reversibly formed acids and supported ruthenium clusters in hot water. Angewandte Chemie, 2007, 119 (40): 7780-7783.

[277] Deng W P, Tan X S, Fang W H, Zhang Q H, Wang Y. Conversion of cellulose into sorbitol over carbon nanotube-supported ruthenium catalyst. Catalysis Letters, 2009, 133 (1-2): 167-174.

[278] Ribeiro L, Órfão J J M, Pereira M F R. Enhanced direct production of sorbitol by cellulose ball-milling. Green Chemistry, 2015, 17 (5): 2973-2980.

[279] Ribeiro L S, Delgado J J, De Melo rfão J J, Ribeiro Pereira M F. Influence of the surface chem-

istry of multiwalled carbon nanotubes on the selective conversion of cellulose into sorbitol. ChemCatChem, 2017, 9 (5): 888-896.

[280] Ribeiro L S, Delgado J J, rfāo J J M, Pereira M F R. Carbon supported Ru-Ni bimetallic catalysts for the enhanced one-pot conversion of cellulose to sorbitol. Applied Catalysis B: Environmental, 2017, 217: 265-274.

[281] Van De Vyver S, Geboers J, Dusselier M, Schepers H, Vosch T, Zhang L A, Van Tendeloo G, Jacobs P A, Sels B F. Selective bifunctional catalytic conversion of cellulose over reshaped Ni particles at the tip of carbon nanofibers. ChemSusChem, 2010, 3 (6): 698-701.

[282] Ding L N, Wang A Q, Zheng M Y, Zhang T. Selective transformation of cellulose into sorbitol by using a bifunctional nickel phosphide catalyst. ChemSusChem, 2010, 3 (7): 818-821.

[283] Van De Vyver S, Geboers J, Schutyser W, Dusselier M, Eloy P, Dornez E, Seo J W, Courtin C M, Gaigneaux E M, Jacobs P A, Sels B F. Tuning the acid/metal balance of carbon nanofiber-supported nickel catalysts for hydrolytic hydrogenation of cellulose. ChemSusChem, 2012, 5 (8): 1549-1558.

[284] Palkovits R, Tajvidi K, Procelewska J, Rinaldi R, Ruppert A. Hydrogenolysis of cellulose combining mineral acids and hydrogenation catalysts. Green Chemistry, 2010, 12 (6): 972-978.

[285] Palkovits R, Tajvidi K, Ruppert A M, Procelewska J. Heteropoly acids as efficient acid catalysts in the one-step conversion of cellulose to sugar alcohols. Chemical Communications, 2011, 47 (1): 576-578.

[286] Hilgert J, Meine N, Rinaldi R, Schuth F. Mechanocatalytic depolymerization of cellulose combined with hydrogenolysis as a highly efficient pathway to sugar alcohols. Energy and Environmental Science, 2013, 6 (1): 92-96.

[287] Geboers J, Van De Vyver S, Carpentier K, Jacobs P, Sels B. Efficient hydrolytic hydrogenation of cellulose in the presence of Ru-loaded zeolites and trace amounts of mineral acid. Chemical Communications, 2011, 47 (19): 5590-5592.

[288] Geboers J, Van De Vyver S, Carpentier K, De Blochouse K, Jacobs P, Sels B. Efficient catalytic conversion of concentrated cellulose feeds to hexitols with heteropoly acids and Ru on carbon. Chemical Communications, 2010, 46 (20): 3577-3579.

[289] Geboers J, Van De Vyver S, Carpentier K, Jacobs P, Sels B. Hydrolytic hydrogenation of cellulose with hydrotreated caesium salts of heteropoly acids and Ru/C. Green Chemistry, 2011, 13 (8): 2167-2174.

[290] Fukuoka A, Dhepe P L. Catalytic conversion of cellulose into sugar alcohols. Angewandte Chemie-International Edition, 2006, 45 (31): 5161-5163.

[291] Han J W, Lee H. Direct conversion of cellulose into sorbitol using dual-functionalized catalysts in neutral aqueous solution. Catalysis Communications, 2012, 19: 115-118.

[292] Zhu W W, Yang H M, Chen J Z, Chen C, Guo L, Gan H M, Zhao X G, Hou Z S. Efficient hydrogenolysis of cellulose into sorbitol catalyzed by a bifunctional catalyst. Green Chemistry, 2014, 16 (3): 1534-1542.

[293] Liu M, Deng W P, Zhang Q H, Wang Y L, Wang Y. Polyoxometalate-supported ruthenium

nanoparticles as bifunctional heterogeneous catalysts for the conversions of cellobiose and cellulose into sorbitol under mild conditions. Chemical Communications, 2011, 47 (34): 9717-9719.

[294] Xi J X, Zhang Y, Xia Q N, Liu X H, Ren J W, Lu G Z, Wang Y Q. Direct conversion of cellulose into sorbitol with high yield by a novel mesoporous niobium phosphate supported Ruthenium bifunctional catalyst. Applied Catalysis A: General, 2013, 459: 52-58.

[295] Liang G F, Cheng H Y, Li W, He L M, Yu Y C, Zhao F Y. Selective conversion of microcrystalline cellulose into hexitols on nickel particles encapsulated within ZSM-5 zeolite. Green Chemistry, 2012, 14 (8): 2146-2149.

[296] Negoi A, Triantafyllidis K, Parvulescu V I, Coman S M. The hydrolytic hydrogenation of cellulose to sorbitol over M (Ru, Ir, Pd, Rh)-BEA-zeolite catalysts. Catalysis Today, 2014, 223: 122-128.

[297] Ji N, Zhang T, Zheng M Y, Wang A Q, Wang H, Wang X D, Chen J G. Direct catalytic conversion of cellulose into ethylene glycol using nickel-promoted tungsten carbide catalysts. Angewandte Chemie-International Edition, 2008, 47 (44): 8510-8513.

[298] Zheng M, Pang J, Sun R, Wang A, Zhang T. Selectivity control for cellulose to diols: dancing on eggs. Acs Catalysis, 2017, 7 (3): 1939-1954.

[299] Zheng M, Pang J, Wang A, Zhang T. One-pot catalytic conversion of cellulose to ethylene glycol and other chemicals: From fundamental discovery to potential commercialization. Chinese Journal of Catalysis, 2014, 35 (5): 602-613.

[300] Delidovich I, Palkovits R. Catalytic isomerization of biomass-derived aldoses: a review. ChemSusChem. 2016, 9 (6): 547-561.

[301] Ji N, Zhang T, Zheng M Y, Wang A Q, Wang H, Wang X D, Shu Y Y, Stottlemyer A L, Chen J G G. Catalytic conversion of cellulose into ethylene glycol over supported carbide catalysts. Catalysis Today, 2009, 147 (2): 77-85.

[302] Li C Z, Zheng M Y, Wang A Q, Zhang T. One-pot catalytic hydrocracking of raw woody biomass into chemicals over supported carbide catalysts: simultaneous conversion of cellulose, hemicellulose and lignin. Energy and Environmental Science. 2012, 5 (4): 6383-6390.

[303] Pang J F, Zheng M Y, Wang A Q, Zhang T. Catalytic hydrogenation of corn stalk to ethylene glycol and 1, 2-propylene glycol. Industrial and Engineering Chemistry Research. 2011, 50 (11): 6601-6608.

[304] Zhao G H, Zheng M Y, Wang A Q, Zhang T. Catalytic conversion of cellulose to ethylene glycol over tungsten phosphide catalysts. Chinese Journal of Catalysis, 2010, 31 (8): 928-932.

[305] Zheng M Y, Wang A Q, Ji N, Pang J F, Wang X D, Zhang T. Transition metal-tungsten bimetallic catalysts for the conversion of cellulose into ethylene glycol. ChemSusChem. 2010, 3 (1): 63-66.

[306] Tai Z J, Zhang J Y, Wang A Q, Zheng M Y, Zhang T. Temperature-controlled phase-transfer catalysis for ethylene glycol production from cellulose. Chemical Communications, 2012, 48 (56): 7052-7054.

[307] Tai Z J, Zhang J Y, Wang A Q, Pang J F, Zheng M Y, Zhang T. Catalytic conversion of cellu-

lose to ethylene glycol over a low-cost binary catalyst of Raney Ni and tungstic acid. ChemSusChem. 2013, 6 (4): 652-658.

[308] Li N X, Zheng Y, Wei L F, Teng H, Zhou J. Metal nanoparticles supported on WO3 nanosheets for highly selective hydrogenolysis of cellulose to ethylene glycol. Green Chemistry, 2017, 19: 682-691.

[309] Ribeiro L S, Órfão J, Órfão J M, Pereira M F R. Hydrolytic hydrogenation of cellulose to ethylene glycol over carbon nanotubes supported Ru-W bimetallic catalysts. Cellulose, 2018, 25 (4): 2259-2272.

[310] Xiao Z, Jin S, Pang M, Liang C. Conversion of highly concentrated cellulose to 1, 2-propanediol and ethylene glycol over highly efficient CuCr catalysts. Green Chemistry, 2013, 15 (4): 891-895.

[311] Pang J, Zheng M, Li X, Jiang Y, Zhao Y, Wang A, Wang J, Wang X, Zhang T. Selective conversion of concentrated glucose to 1, 2-propylene glycol and ethylene glycol by using RuSn/AC catalysts. Applied Catalysis B: Environmental, 2018, 239: 300-308.

[312] Ferrini P, Dijkmans J, Clercq R D, Vyver S V D, Dusselier M, Jacobs P A, Sels B F. Lewis acid catalysis on single site Sn centers incorporated into silica hosts. Coordination Chemistry Reviews, 2017, 343: 220-255.

[313] Hensen E J M, Pidko E A, Degirmenci V, Van Santen R A. Glucose activation by transient Cr^{2+} dimers. Angewandte Chemie-International Edition, 2010, 49 (14): 2530-2534.

[314] Choudhary V, Pinar A B, Lobo R F, Vlachos D G, Sandler S I. Comparison of homogeneous and heterogeneous catalysts for glucose-to-fructose isomerization in aqueous media. ChemSusChem, 2013, 6 (12): 2369-2376.

[315] Liu C, Carraher J M, Swedberg J L, Herndon C R, Fleitman C N, Tessonnier J-P. Selective base-catalyzed isomerization of glucose to fructose. Acs Catalysis, 2014, 4 (12): 4295-4298.

[316] Carraher J M, Fleitman C N, Tessonnier J P. Kinetic and mechanistic study of glucose isomerization using homogeneous organic brønsted base catalysts in water. Acs Catalysis. 2015, 5 (6): 3162-3173.

[317] Nguyen H, Nikolakis V, Vlachos D G. Mechanistic insights into Lewis acid metal salt-catalyzed glucose chemistry in aqueous solution. Acs Catalysis, 2016, 6 (3): 1497-1504.

[318] Tang J Q, Guo X W, Zhu L F, Hu C W. Mechanistic Study of Glucose-to-Fructose Isomerization in Water Catalyzed by [Al(OH)$_2$(aq)]$^+$. Acs Catalysis, 2015, 5 (9): 5097-5103.

[319] Bermejo-Deval R, Orazov M, Gounder R, Hwang S J, Davis M E. Active sites in Sn-β for glucose isomerization to fructose and epimerization to mannose. Acs Catalysis, 2014, 4 (7): 2288-2297.

[320] Li G N, Pidko E A, Hensen E J M. Synergy between Lewis acid sites and hydroxyl groups for the isomerization of glucose to fructose over Sn-containing zeolites: a theoretical perspective. Catalysis Science and Technology, 2014, 4 (8): 2241-2250.

[321] Rai N, Caratzoulas S, Vlachos D G. Role of silanol group in Sn-Beta zeolite for glucose isomerization and epimerization reactions. Acs Catalysis, 2013, 3 (10): 2294-2298.

[322] Roman-Leshkov Y, Moliner M, Labinger J A, Davis M E. Mechanism of glucose isomerization

using a solid Lewis acid catalyst in water. Angewandte Chemie-International Edition, 2010, 49 (47): 8954-8957.

[323] Bermejo-Deval R, Assary R S, Nikolla E, Moliner M, Roman-Leshkov Y, Hwang S J, Palsdottir A, Silverman D, Lobo R F, Curtiss L A, Davis M E. Metalloenzyme-like catalyzed isomerizations of sugars by Lewis acid zeolites. Proceedings of the National Academy of Sciences of the United States of America, 2012, 109 (25): 9727-9732.

[324] Yang G, Pidko E A, Hensen E J M. The mechanism of glucose isomerization to fructose over Sn-BEA zeolite: a periodic density functional theory study. ChemSusChem. 2013, 6 (9): 1688-1696.

第三章

两相体系中转化纤维素制备5-羟甲基糠醛

3.1 引言

己糖脱水所生成的5-羟甲基糠醛（HMF）是一种非常重要的木质纤维素衍生化学品，被认为是生产生物质衍生液体燃料和其他重要化学品的关键物质。一方面，HMF可以通过后续反应制取乙酰丙酸、2,5-呋喃二羧酸、2,5-二甲酰呋喃、二甲基呋喃和二羟甲基呋喃等下游化工产品；另一方面，HMF可以通过加氢脱氧、碳链增长等手段得到二甲基呋喃、长链液体烷烃等液体燃料。然而，由于木质纤维素本身结构的顽抗性和HMF在水溶剂中的不稳定性，导致利用木质纤维素高效制备HMF成为一个巨大挑战。

如第二章2.3节所述，在酸性水溶液中转化纤维素制备HMF通常收率都不高，因此，人们开发了以离子液体、极性有机溶剂和双相溶剂作为反应介质去转化碳水化合物制备HMF的方法。然而，离子液体和极性有机溶剂在用于大规模工业化转化木质纤维素生产HMF上存在许多难以克服的问题，比如成本过高、产物分离能耗高，且分离后的HMF产物中会存在部分N、S等杂原子而不适宜于生产液体燃料。因此，在本章的研究中采用由水与有机溶剂组成的两相体系去转化纤维素。两相体系的优势在于，可以采用沸点较低的有机溶剂作为萃取相，这样能够较容易地采用蒸馏的方法将HMF从反应溶剂中分离出来。此外，低沸点有机溶剂通常只含有C、H、O三种元素，即使分离出的HMF中含有少量萃取溶剂，也不影响其后续作为生产生物质液体燃料的原料。特别地，丁醇、四氢呋喃（THF）、甲基四氢呋喃、γ-戊内酯、烷基酚等有机溶剂都能够从生物质中得到，而它们也都被证明能够用作两相体系中的萃取相[1-3]。在多种用于两相体系萃取相的有机溶剂中，四氢呋喃被认为具有较好的性能且其沸点较低，易于从反应产物中蒸馏出来进行重复使用。Yang等报道在水-THF组成的两相体系中采用$AlCl_3$催化纤维素转化得到37%的HMF[4]。

用于转化纤维素的催化剂包括无机酸、金属盐和固体酸。三类催化剂中，无机酸对纤维素水解的催化作用最佳，但是无机酸用在两相体系中可能会被萃取到有机相，导致催化剂在回收过程中出现大量损失。金属盐通常在有机溶剂中的溶解度较低，所以较适宜用于两相体系，但是金属盐对纤维素水解的催化作用又相对较弱。相比之下，多元无机酸的酸式盐如硫酸氢钠、硫酸氢钾则同时具备了无机酸的强酸性和金属盐难溶于有机溶剂的特性，因此似乎更加适宜于用作两相体系中纤维素转化的催化剂。

本章主要介绍了在由水-四氢呋喃组成的两相体系中利用硫酸氢钠和硫酸

锌协同催化纤维素制备 HMF，并且得到了 53% 的 HMF 收率。研究发现，在由水-四氢呋喃组成的反应体系中，水和四氢呋喃之间是互溶的，但无机盐的添加会导致少量的水相从有机相中分离出来形成两相体系。该两相体系中，纤维素的亲水性较强使得纤维素会优先与水相接触，但是由于水相体积较小，所以水相会附着在纤维素的表面形成一层液膜。然后，在搅拌的状态下，覆盖了水相液膜的纤维素分散于有机相中，形成类微乳反应体系。在这种类微乳反应体系中，有机相是主要的分散介质，而水相则主要作为反应发生的场所。由于大量的水溶解到四氢呋喃中，只有少量的水在无机盐的作用下析出，所以水相中无机盐的浓度较高，从而高效催化纤维素转化为 HMF。在水相中生成的 HMF 则迅速被萃取到有机相中并得到保护，避免了 HMF 的降解反应。

3.2 转化纤维素制备 5-羟甲基糠醛的过程及产物的分析表征

（1）纤维素制备 HMF 的过程

催化转化纤维素为 HMF 的反应在一个材质为 316L 的、容积为 100mL 的机械搅拌反应釜中进行。在一个典型的实验中，1g 纤维素、4mL 去离子水、40mL 四氢呋喃和一定量的催化剂都加入反应器中。将高纯氮气通入反应器中置换出反应器中的空气，并将反应器压力增加至 3.0MPa。然后，在搅拌状态下将反应器升温到目标温度并且保持预设的时间。反应结束后，冷却反应釜至室温，固体残渣与液体产物通过 0.45μm 的滤膜进行过滤，得到的水相和有机相液体经过稀释后，用气相色谱-质谱联用技术（GC-MS）和液相色谱技术（HPLC）对液体产物进行定性和定量分析。收集到的固体残渣用去离子水洗涤，并且在 105℃下干燥 3h，并用热重（TG）、扫描电子显微镜（SEM）、元素分析和傅里叶变换红外光谱（FT-IR）进行分析。

（2）液体产物的分析

对于 HMF 和乙酰丙酸的定性分析在 GC-MS（HP5890；MSD，HP5972A）上进行。色谱柱为 DB-5MS 毛细管柱（30m×0.25mm×0.25μm）。色谱柱柱温程序为：在 40℃保持 4min，然后以 10℃/min 的升温速率升温至 250℃，并在 250℃保持 3min。

纤维二糖、葡萄糖、果糖、HMF、糠醛和乙酰丙酸在反应液体中的浓度通过 HPLC 进行定量。色谱柱为 AminexHPX-87H 有机酸柱，操作柱温

50℃。用 Waters 410 折光检测器（操作温度 50℃）去检测纤维二糖、葡萄糖和乙酰丙酸，用紫外检测器（操作波长 284nm）去检测糠醛和 HMF。流动相是 5mmol/L 的稀硫酸水溶液，流速为 0.55mL/min。产物的定量采用外标法。液体样品在进行 HPLC 分析前都用去离子水稀释 50 倍。

产物的理论收率通过下面的方程进行计算，假设纤维素中的葡萄糖苷单元的摩尔质量为 162g/mol，则

$$\text{收率} = \frac{(\text{有机相中浓度} \times \text{有机相体积} + \text{水相中浓度} \times \text{水相体积})/M_i}{\text{纤维素质量}/162} \times 100\%$$

式中，M_i 是 HMF、糠醛和乙酰丙酸每种物质的摩尔质量，分别为 126g/mol、96g/mol 和 116g/mol。

（3）固体产物的表征

对反应后固体残渣的 TG 分析采用 NETZSCH-STA 409 PC DSC-SP 热重仪。将 5～10mg 样品置于反应器中，并且在空气气氛下将温度从 40℃以 10℃/min 的速率升高到 500℃。

纤维素及残渣的形貌通过 SEM（HITACHI S-4800）进行分析。残渣的元素分析在 VarioELZCHN 元素分析仪上面进行。操作条件为：0.12MPa 的氦和 0.2MPa 的氧气气氛下，在 1145℃下采用 TCD 检测器进行检测。

纤维素及残渣的结构通过 FT-IR 进行分析，扫描波长为 400～4000cm^{-1}。样品在分析前先与 KBr 混合研磨，并压成薄片。

3.3 在两相体系中金属盐催化剂的筛选

首先考察了在由四氢呋喃与水按 10∶1 的比例组成的反应体系中，几种常见的金属盐 [$Fe_2(SO_4)_3$、$Al_2(SO_4)_3$、$ZnSO_4$、$NaHSO_4$、$FeCl_3$、$ZnCl_2$、$AlCl_3$] 对纤维素降解的催化作用（图 3-1）。

在不加入催化剂的情况下，在 160℃反应 1h 后，得到固体残渣收率达到 95%，而生成的 HMF 及乙酰丙酸都几乎可以忽略。反应后的残渣主要是未转化的纤维素，表明纤维素的转化率不到 5%。两种锌盐（$ZnSO_4$、$ZnCl_2$）对纤维素转化的催化效果也很差。如图 3-1 所示，添加 $ZnSO_4$ 和 $ZnCl_2$ 之后，80%以上的纤维素也都未转化，且生成的 HMF 也几乎可以忽略。相比于锌盐，铝盐 [$AlCl_3$ 和 $Al_2(SO_4)_3$] 和铁盐 [$Fe_2(SO_4)_3$ 与 $FeCl_3$] 对纤维素的降解具有明显的催化效果。$AlCl_3$ 对纤维素转化为 HMF 的催化作用已经被四川大学胡常伟等报道过[4]，在这里利用 $AlCl_3$ 转化纤维素得到 32%的 HMF，

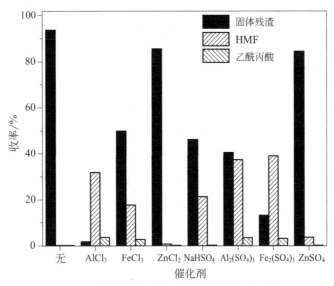

图 3-1 不同催化剂对纤维素转化的催化效果

反应条件：1.0g 纤维素，40mL THF，4mL H_2O，160℃，60min，0.5mmol 催化剂

与他们的结果相近。其他文献也已经报道了 $FeCl_3$ 在水溶剂中能够催化纤维素转化生成乙酰丙酸[5,6]，而 $Al_2(SO_4)_3$ 能够在醇溶剂中转化纤维素生成乙酰丙酸酯[7]，实际上也就表明这些催化剂能转化葡萄糖异构化反应和果糖脱水反应。因此，这两种金属盐能够转化纤维素生成 HMF 是意料之中的。除了 $AlCl_3$ 之外，几乎所有的盐对纤维素降解的催化效果均与 HMF 收率成正相关。使用 $Al_2(SO_4)_3$ 和 $Fe_2(SO_4)_3$ 作为催化剂时，得到的 HMF 收率甚至高于使用 $AlCl_3$ 时得到的 HMF 收率，尽管反应剩余更多的未转化纤维素。采用各类盐作为催化剂时，乙酰丙酸的收率都很低，表明在反应体系中 HMF 的水合反应被有效抑制[4]。如果对反应参数进行优化，这些催化剂催化纤维素转化为 HMF 的收率还可以继续提高。

目前在多数制备 HMF 的研究中，特别是在离子液体中，人们都倾向于使用氯盐作为催化剂[2,8]。这是因为人们认为氯离子可以与葡萄糖的羟基产生氢键，从而促进葡萄糖与果糖之间的脱水反应。仅有少量文献研究了硫酸盐对葡萄糖转化的影响[9]。上面的研究发现，$Al_2(SO_4)_3$ 和 $Fe_2(SO_4)_3$ 能够催化纤维素转化 HMF，表明硫酸盐对转化木质纤维素制备 HMF 同样具有较好的作用。因为硫酸根可以通过与 $Ca(OH)_2$ 反应生成沉淀而除去[10]，而溶液中的氯离子不仅难以除去，而且易导致金属反应设备的晶间腐蚀，因此，笔者认为利用硫酸盐作催化剂比利用金属氯盐作为催化剂更具有优势。

3.4 NaHSO$_4$-ZnSO$_4$ 协同催化纤维素降解制备 5-羟甲基糠醛

NaHSO$_4$ 是一种常见的酸式盐，属于中强酸，能够电离出氢离子从而能够有效催化纤维素的水解，又属于盐，从而在两相体系中具有盐析效应。上面的结果表明 NaHSO$_4$ 在两相体系中对转化纤维素具有较好的效果。另外，因为大量文献指出锌离子对葡萄糖异构化反应具有催化作用[9,11]，而纤维素降解为 HMF 涉及纤维素水解为葡萄糖、葡萄糖异构化为果糖及果糖脱水三个步骤，因此，笔者采用 NaHSO$_4$ 与 ZnSO$_4$ 共同催化葡萄糖水解制备 HMF。

在上述的研究中，在仅使用 NaHSO$_4$ 时，即可以得到 20% 的 HMF 收率。实际上，通过优化 NaHSO$_4$ 的用量（表 3-1 中序列 3～5），可以提高 HMF 收率到约 40%，表明 NaHSO$_4$ 中的氢离子对于纤维素制备 HMF 中涉及的三个反应（纤维素的水解、葡萄糖的异构化和果糖的脱水）都具有较好的催化效果。然而，相比于仅使用 NaHSO$_4$ 时得到的 HMF 收率，ZnSO$_4$ 的添加明显地提高了 HMF 的收率（表 3-1 中序列 6～10）。当 NaHSO$_4$ 和 ZnSO$_4$ 用量分别为 1.8mmol 和 2.8mmol 时，HMF 的收率可以提高到 53.2%。在这里得到的 HMF 的收率与在离子液体中用 CrCl$_3$ 作催化剂时得到的收率（62%）接近[12]，并且远高于在两相体系中采用 AlCl$_3$ 作为催化剂时得到的收率（37%）[4]。

由于水与四氢呋喃是互溶的，所以在反应体系中不添加盐作催化剂时，反应体系为单相。盐的添加使得水相从四氢呋喃中分离出来，得到溶解了金属盐的水相和未溶解金属盐的四氢呋喃相。表 3-1 中列出了水相的体积随催化剂用量的变化情况。可以看到，在仅添加 NaHSO$_4$ 时，水相的体积始终非常少（加入 1.8mmol NaHSO$_4$ 时水相体积也仅 0.3mL），而加入 ZnSO$_4$ 之后，水相的体积从 0.3mL 增加到 1.5mL，表明硫酸锌的添加除了起到促进葡萄糖异构化为果糖的作用外，还起到增加水相体积、调节水相中 NaHSO$_4$ 浓度的作用。

表 3-1 NaHSO$_4$-ZnSO$_4$ 体系对直接转化纤维素为 HMF 的催化效果

序列	催化剂/mmol		水相体积/mL	产物收率/%	
	NaHSO$_4$	ZnSO$_4$		HMF	乙酰丙酸
1	—	—	0	<0.5	<0.5
2	—	2.8	1.0	0.7	<0.5
3	0.9	—	0.2	29.8	<0.5

续表

序列	催化剂/mmol		水相体积/mL	产物收率/%	
	NaHSO$_4$	ZnSO$_4$		HMF	乙酰丙酸
4	1.8	—	0.3	36.7	<0.5
5	2.7	—	0.4	28.4	10.4
6	1.8	0.5	0.5	45.7	<0.5
7	1.8	1.0	0.8	48.7	<0.5
8	1.8	2.8	1.2	53.2	<0.5
9	1.8	3.5	1.4	48.1	<0.5
10	1.8	4.2	1.5	47.8	0.7
11	2.3	2.8	1.5	47.4	6.8
12	2.7	2.8	1.5	45.9	12.9

3.4.1 反应时间的影响

在优化的催化剂用量下，考察了 HMF、葡萄糖和固体残渣的收率随反应时间的变化情况（图 3-2）。可以看到，在 30min 内固体残渣的收率下降到 25%，而 HMF 的收率则达到 35%，表明纤维素已经大量发生了解聚且被转化成了 HMF。在 50min 时，最低的残渣收率为 8.0%，表明超过 92% 的纤维素在 50min 内被转化。在 50min 之后，固体残渣的收率开始上升，这是由于生成了固体胡敏素。HMF 的收率在 60min 时达到最大值，之后发生缓慢下降表明 HMF 在反应体系中仍然会发生降解。

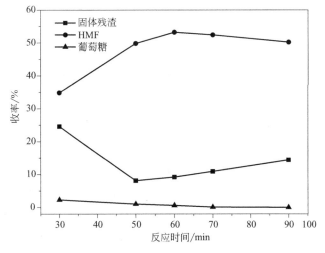

图 3-2　HMF、葡萄糖和固体残渣的收率随反应时间的变化情况

在整个反应过程中，只有少量的葡萄糖被检测到，表明反应体系对葡萄糖的转化非常迅速，刚刚生成的葡萄糖立即发生了异构化及脱水反应，所以葡萄糖浓度在反应体系中维持在一个较低水平。反应体系中葡萄糖的低浓度是反应体系能够得到较高的HMF收率的原因之一。因为纤维素转化为HMF涉及纤维素水解和葡萄糖脱水两个过程，而其中纤维素水解的活化能（大约180kJ/mol）远高于葡萄糖降解的活化能（大约135kJ/mol）[13-15]，所以水解纤维素是其中的速控步。在这个反应过程中，当纤维素被水解之后，生成的葡萄糖被水相中存在的大量的H^+和Zn^{2+}迅速转化掉，从而保证葡萄糖在水相中的浓度较低，从而有效地抑制了葡萄糖与其他物质之间的聚合反应。

表3-2对比了纤维素以及在不同时间下得到的固体残渣的元素组成。可以看到，残渣中含有大约61.6%～63.6%（质量分数）的碳，4.8%～4.9%（质量分数）的氢和约30%（质量分数）的氧，与文献报道的胡敏素的元素组成接近[16,17]。残渣中的碳含量远高于纤维素中的碳含量，氧含量则远低于纤维素中的氧含量，表明残渣中主要纤维素经过脱水/碳化得到胡敏素。与在60min时得到的残渣相比，在120min时得到的残渣的碳元素含量更高而氧元素含量更低，表明生成的残渣可能继续发生了脱水/碳化反应。

表3-2 纤维素及固体残渣的元素组成

材料	元素组成(质量分数)/%		
	C	H	O
纤维素	44.1	5.9	50.0
残渣(60min)	61.6	4.9	31.8
残渣(120min)	63.6	4.8	29.9

将透明的有机相添加到去离子水中会形成一些褐色沉淀，表明在反应过程中生成了某些能溶于THF但不溶于水的聚合物。对这类不溶于水的聚合物进行红外分析（图3-3），可以看到这些分离出的可溶性聚合物中主要包含羟基、羰基、呋喃环（或苯环）、C—C单键、C—O单键等官能团，与文献所报道的胡敏素的结果相一致[16-18]。

纤维素、固体残渣、胡敏素和焦炭的SEM照片如图3-4所示。纤维素是表面致密的粒径50μm的颗粒，而反应后得到的固体残渣是表面松散多孔的、不规则条状物质，这表明纤维素在转化过程中其表面被催化剂强烈地腐蚀了。THF可溶性胡敏素是1～3μm的球状，表明这些物质是某些可溶性物质聚合而成的副产物。

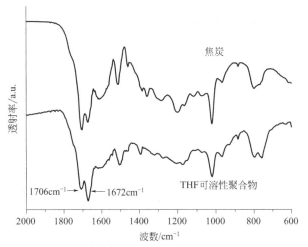

图 3-3 THF 可溶性聚合物与葡萄糖水解得到的焦炭的红外谱图

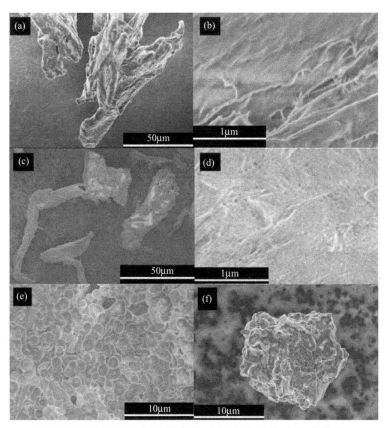

图 3-4 纤维素 [（a）和（b）]、固体残渣 [（c）和（d）]、
胡敏素（e）和焦炭（f）的 SEM 照片

3.4.2 反应温度对纤维素转化的影响

在 140~180℃ 的温度区间内，反应温度对纤维素转化的影响如图 3-5 所示。在 140℃ 的较低温度下，仅有 35.8% 的 HMF 生成，且有 41.6% 的纤维素未发生反应。提高温度能明显地促进纤维素的转化并提高 HMF 的收率。温度从 140℃ 提高到 160℃，纤维素的转化率从约 60% 提高到超过 90%，HMF 的收率也提高到了 53.2%。然而，当温度升高到 180℃ 时，HMF 收率下降到 45% 并生成约 10% 的乙酰丙酸。这可能是由于在过高的温度下，被萃取到有机相中的 HMF 仍然会重新溶解到水相中并发生水合反应。

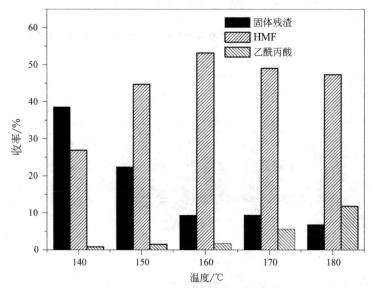

图 3-5 反应温度对纤维素转化的影响

3.4.3 水/有机相的体积比对反应的影响

研究发现，反应体系中水的用量会极大地影响 HMF 的收率（表 3-3）。当水的添加量为 1mL 时，反应后收集到的水相体积仅有 0.2mL，此时转化纤维素能得到 37.7% 的 HMF 收率和 10.3% 的固体残渣。随着水的添加量的增加，反应体系中的水相体积也明显增大了，而 HMF 的收率开始增大。当水的用量增加到 4mL 时，反应后收集到的水相体积达到了 1.2mL，有机相体积则上升到 43.2mL，而 HMF 的收率也上升到了 53.2%。然而，当水的用量增加到 8mL 时，只有 1.7mL 的水相被检测到，而 HMF 的收率下降到 41.8%。特别地，当水的用量为 16mL 时，反应后收集到的水相体积仅为 1.4mL，而 HMF 的收率则迅速下降到 8.2%，并且固体残渣达到了 85.6%。

表 3-3 反应体系中水的用量对 HMF 收率的影响

水用量 /mL	催化剂状态	溶剂体积/mL		HMF 收率 /%	固体残渣 收率/%
		水相	有机相		
1	部分溶解	0.2	40.8	37.7	10.3
2	部分溶解	0.5	41.4	47.6	9.7
4	部分溶解	1.2	43.2	53.2	8.1
8	完全溶解	1.7	46.3	41.8	17.2
16	完全溶解	1.4	55.1	8.2	85.6

注：反应条件为 1g 纤维素，40mL THF，1.8mmol $NaHSO_4$ 和 2.8mmol $ZnSO_4$，160℃，60min。

可以看到，随着反应体系中水用量的增加，固体残渣的收率呈先下降后显著上升的趋势，而 HMF 的收率呈现出先上升后下降的趋势。这可以通过催化剂的浓度和水相体积的变化来进行解释。在水用量为 1mL 时，反应后收集到的水相体积仅 0.2mL，且水相中的催化剂处于部分溶解状态。在这种情况下，一方面由于水相体积过少，而水相中催化剂的浓度又过高，导致水相不能充分地与纤维素表面接触而转化纤维素，另一方面与纤维素接触的催化剂浓度过高导致 HMF 易水解开环转化为固体胡敏素。因此，在这种情况下固体残渣较多而 HMF 收率较低。随着水用量从 1mL 增加到 4mL，反应体系中的水相体积显著增大，于是水相中的催化剂浓度逐渐下降，这两个变化都能够使得纤维素能够较好地水解并转化生成 HMF。当催化剂浓度下降到某一合理值而水相体积增大到某一合理值时，HMF 的收率达到最高值。当继续增加水用量到 8mL 和 16mL 时，水相体积并没有非常显著地增大，表明大量的水溶解到水相中。这种情况下，由于有机相中溶解了大量的水，使得有机相逐渐变得对无机盐具有一定的溶解度，所以部分催化剂溶解到有机相中，从而导致水相中的催化剂浓度大大下降，进而导致催化剂对纤维素水解的催化作用显著下降，最终导致 HMF 收率的下降和固体残渣收率的上升。

3.4.4 反应体系对其他可溶性糖的转化

上述研究指出，葡萄糖在水相体系中的浓度极低，从而抑制葡萄糖及其降解产物的聚合反应，可能是反应体系能够将纤维素高效转化为 HMF 的主要原因。因此，笔者研究了反应体系对葡萄糖转化为 HMF 的催化性能。可以看到，当反应体系采用葡萄糖为原料时，仅能够生成 35.7% 的 HMF，却生成了

21.7%的固体残渣。这表明，反应体系对转化葡萄糖制备HMF的效果较差。与葡萄糖相同，纤维二糖和淀粉在反应体系中也同样产生了约25%的固体残渣，而生成的HMF则均低于20%（表3-4）。SEM图片表明它是大于10μm的不规则颗粒，与未反应的纤维素和球状的可溶性胡敏素不同（图3-4）。FT-IR光谱显示，固体残渣与可溶性胡敏素的红外非常相似，在1604cm^{-1}、1510cm^{-1}和1395cm^{-1}处的吸收峰表明在两种固体中都存在呋喃环[19]。然而，与可溶性胡敏素不同，从可溶性糖降解得到的固体残渣在1706cm^{-1}处（与C=C键不共轭的羰基）的吸收峰强度高于在1672cm^{-1}（与C=C键共轭的羰基）处的强度，表明此处生成的固体残渣中含有更少的C=O键与C=C键共轭[20]。

通常情况下，焦炭是在一些比较苛刻的水热条件下形成的，比如高温、高的酸浓度、高的糖浓度等[20-22]。Srokol等在研究葡萄糖的降解时发现，在低的葡萄糖浓度时几乎没有焦炭生成，在较高的葡萄糖浓度下则生成大量的焦炭残渣[22]。在这里，焦炭的形成很可能是由高浓度的糖溶液所导致的。由于葡萄糖、纤维二糖、淀粉都只能溶解在水中而不能溶解在THF中，因此，当采用可溶性糖作为反应原料时，由于水相的体积较少，所以在水相中这些糖的浓度高达40%。如此高浓度的糖在高温条件下极易发生聚合反应而生成焦炭。

为了进一步证明水热焦炭的生成是由可溶性糖在水相中的浓度过高所致，笔者考察了葡萄糖浓度对HMF和固体焦炭生成的影响（表3-4）。可以看到，提高葡萄糖的用量到5.0g会导致HMF收率下降到仅24.1%而固体残渣的质量分数提高到36.3%，而降低葡萄糖的用量可以显著提高反应体系对催化葡萄糖生成HMF的选择性。当葡萄糖的用量为0.2g时，HMF收率可以提高到45.2%，而固体残渣的收率则下降到10.9%。这一结果充分证明，可溶性糖在反应体系中的浓度会大大影响HMF的收率。

表3-4 反应体系对葡萄糖、淀粉和纤维素转化的比较

原料	原料用量/g	时间/min	转化率/%	产物收率/%			固体残渣收率/%
				HMF	糠醛	乙酰丙酸	
葡萄糖	1.0	30	98.4	35.7	0.5	<0.5	21.7
淀粉	1.0	45	100	16.9	0.7	<0.5	24.1
纤维二糖	1.0	45	100	17.6	1.5	<0.5	23.8
葡萄糖	5.0	30	95.5	24.1	0.6	<0.5	36.3
葡萄糖	0.5	30	99.1	40.9	0.6	<0.5	14.3
葡萄糖	0.2	30	100	45.2	<0.5	<0.5	10.9

3.4.5 原料用量对 5-羟甲基糠醛收率的影响

提高反应原料的用量能够提高 HMF 在反应溶液中的浓度，从而降低 HMF 的分离能耗。因此，笔者继续研究了纤维素用量对 HMF 收率的影响。如表 3-5 所示，当纤维素的浓度提高到 11% 时，HMF 的收率仍然能够保持在较高的 42.5%，表明系统仍然能够有效地催化纤维素的转化。继续提高纤维素的用量，HMF 的收率开始下降，而固体残渣的质量开始增大，表明提高纤维素的浓度会导致葡萄糖及其降解产物的聚合/缩聚等反应的增多，从而生成更多的固体聚合物[23]。当纤维素用量增加到 27% 时，HMF 浓度可以提高到 5.9%，但是 HMF 的收率已经下降到仅 22.6%。可见，在得到高浓度 HMF 的同时并且保持 HMF 的高收率仍然是一个需要解决的难题。

表 3-5 纤维素用量对 HMF 收率的影响

纤维素浓度（质量分数）/%	时间/min	产物收率/%		HMF 浓度/%	固体残渣收率/%
		HMF	LA		
11.0	60	42.5	<0.5	3.3	12.2
19.8	60	35.0	1.4	5.0	17.1
21.3	60	33.5	1.9	5.2	19.4
22.8	60	31.8	3.3	5.4	21.2
24.2	60	24.9	4.8	5.5	25.4
27.0	90	22.6	5.7	5.9	30.7

3.5 反应体系分析及液膜催化概念的提出

如前所述，在水-四氢呋喃组成的两相体系中，金属盐催化剂主要存在于水相，而产物主要存在于有机相。因此笔者认为，催化剂在水相中的富集效应是导致反应体系对纤维素转化为 HMF 具有优异效果的主要原因。当反应体系为 4mL 水、1.8mmol $NaHSO_4$、2.8mmol $ZnSO_4$ 和 40mL 的四氢呋喃时，仅得到 1.2mL 的水相和 43.2mL 的有机相，表明在水相中 $NaHSO_4$ 和 $ZnSO_4$ 的浓度分别为 1.5mol/L 和 2.3mol/L。如此高浓度的催化剂对于纤维素的水解及葡萄糖的异构化和脱水都具有很好的催化效果。Dumesic 等研究硫酸催化纤维素在由水与 γ-戊内酯组成的体系中转化生产乙酰丙酸时，也发现了类似的催化剂富集效应。他们发现，在催化剂用量相同的情况下，当γ-戊内酯与水按9∶1的比例作为反应介质时，反应体系对纤维素水解的催化作用远远高于仅以水为溶剂时的效果[24-26]。这可能是由于在γ-戊内酯与水组成的溶剂中转化纤维素时，纤维素会优先与水亲和，而硫酸的极性较强，也易溶解在水中，使得在纤维素表面形成一层硫酸浓度较高的液膜催化层。当只采用水作为催化剂时，硫酸均匀地分散于整个反应溶剂中导致其浓度相对较低，所以对纤维素水解的催化作用较弱。

有机相与水相的高体积比是该反应体系能够有效转化纤维素为糠醛的另一个主要原因。如表3-3所示，虽然反应体系中四氢呋喃和水加入的体积比为10∶1，但是由于部分水会溶解于四氢呋喃，所以实际上有机相的体积是水相体积的25倍以上。因此，在剧烈搅拌的情况下，含有催化剂的水相和纤维素颗粒都会分散在有机相中。另一方面，纤维素的表面分布有大量的羟基，对含有羟基的物质（特别是水）具有很强的亲和力[27]。Khazraji等指出，即使在室温条件下，当纤维素暴露在空气中时，其表面至少覆盖有一层水分子[27]。因此在两相体系中，纤维素的表面应该优先与水相接触而不是与有机相接触，从而包覆在纤维素表面形成一层溶解了催化剂的酸性水膜（图3-6）。假设纤维素的表面完全被水相所包覆，因为纤维素的比表面积约为 $0.7 m^2/g$，而水相的体积是 $1.2 mL$，可以计算出纤维素表面的水层厚度为 $1.7 \mu m$，远远低于纤维素的直径。所以完全可以认为，这层含有高浓度催化剂的薄水膜高效地催化纤维素的水解反应和葡萄糖的脱水反应，而形成的 HMF 分子快速地转移到

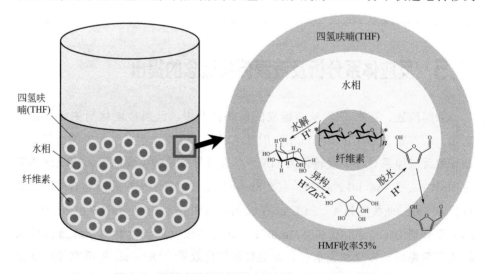

图 3-6　酸性水膜催化纤维素降解为 HMF 的示意图

有机相中去避免其后续降解。在这里,有机相不仅作为 HMF 的保护溶剂,而且作为水相和纤维素的分散剂,而水作为反应物去水解纤维素以及作为溶剂去溶解催化剂。

3.6 结论

本章开发了一种以 $NaHSO_4$ 和 $ZnSO_4$ 为催化剂、以水和四氢呋喃组成的两相体系为反应介质的将纤维素转化为 HMF 的催化反应方法。在优化催化剂用量和反应条件的基础上,得到了 53% 的 HMF 收率,以及 96% 的纤维素转化率。催化剂在水相中的富集所导致的高浓度催化剂以及有机相与水相的高体积比是反应体系具有较好效果的主要因素。在此反应体系中,四氢呋喃是主要溶剂,尽管它可能带来价格昂贵等一系列不利因素,但是四氢呋喃的沸点较低,从而能够降低 HMF 分离所需的能耗。总之,该反应体系有望应用于工业化利用木质纤维素生产 HMF。

参 考 文 献

[1] Roman-Leshkov Y, Dumesic J A. Solvent effects on fructose dehydration to 5-hydroxymethylfurfural in biphasic systems saturated with inorganic salts. Topics in Catalysis, 2009, 52 (3): 297-303.

[2] Pagan-Torres Y J, Wang T F, Gallo J M R, Shanks B H, Dumesic J A. Production of 5-hydroxymethylfurfural from glucose using a combination of Lewis and Brønsted acid catalysts in water in a biphasic reactor with an alkylphenol solvent. Acs Catalysis, 2012, 2 (6): 930-934.

[3] Alonso D M, Wettstein S G, Bond J Q, Root T W, Dumesic J A. Production of biofuels from cellulose and corn stover using alkylphenol solvents. ChemSusChem. 2011, 4 (8): 1078-1081.

[4] Yang Y, Hu C W, Abu-Omar M M. Conversion of carbohydrates and lignocellulosic biomass into 5-hydroxymethylfurfural using $AlCl_3 \cdot 6H_2O$ catalyst in a biphasic solvent system. Green Chemistry, 2012, 14 (2): 509-513.

[5] Peng L C, Lin L, Zhang J H, Zhuang J P, Zhang B X, Gong Y. Catalytic conversion of cellulose to levulinic acid by metal chlorides. Molecules, 2010, 15 (8): 5258-5272.

[6] Zheng X, Zhi Z, Gu X, Li X, Zhang R, Lu X. Kinetic study of levulinic acid production from corn stalk at mild temperature using $FeCl_3$ as catalyst. Fuel, 2017, 187: 261-267.

[7] Huang Y B, Yang T, Lin Y T, Zhu Y Z, Li L C, Pan H. Facile and high-yield synthesis of methyl levulinate from cellulose. Green Chemistry, 2018, (20): 1323-1334

[8] Zhao H B, Holladay J E, Brown H, Zhang Z C. Metal chlorides in ionic liquid solvents convert sugars to 5-hydroxymethylfurfural. Science, 2007, 316 (5831): 1597-1600.

[9] Heeres H J, Rasrendra C B, Makertihartha I G B N, Adisasmito S. Green chemicals from D-glu-

cose: systematic studies on catalytic effects of inorganic salts on the chemo-selectivity and yield in aqueous solutions. Topics in Catalysis, 2010, 53 (15-18): 1241-1247.

[10] Li W Z, Xu H, Wang J, Yan Y J, Zhu X F, Chen M Q, Tan Z C. Studies of monosaccharide production through lignocellulosic waste hydrolysis using double acids. Energy and Fuels, 2008, 22 (3): 2015-2021.

[11] Nagorski R W, Richard J P. Mechanistic imperatives for aldose-ketose isomerization in water: specific, general base-and metal ion-catalyzed isomerization of glyceraldehyde with proton and hydride transfer. Journal of the American Chemical Society, 2001, 123 (5): 794-802.

[12] Yu H B, Wang P, Zhan S H, Wang S Q. Catalytic hydrolysis of lignocellulosic biomass into 5-hydroxymethylfurfural in ionic liquid. Bioresource Technology, 2011, 102 (5): 4179-4183.

[13] Pilath H M, Nimlos M R, Mittal A, Himmel M E, Johnson D K. Glucose reversion reaction kinetics. Journal of Agricultural and Food Chemistry, 2010, 58 (10): 6131-6140.

[14] Saeman J F. Kinetics of wood saccharification-hydrolysis of cellulose and decomposition of sugars in dilute acid at high temperature. Industrial and Engineering Chemistry, 1945, 37 (1): 43-52.

[15] Rasrendra C B, Soetedjo J N M, Makertihartha I G B N, Adisasmito S, Heeres H J. The Catalytic conversion of D-glucose to 5-hydroxymethylfurfural in DMSO using metal salts. Topics in Catalysis, 2012, 55 (7-10): 543-549.

[16] Sumerskii I V, Krutov S M, Zarubin M Y. Humin-like substances formed under the conditions of industrial hydrolysis of wood. Russian Journal of Applied Chemistry, 2010, 83 (2): 320-327.

[17] Patil S K R, Lund C R F. Formation and growth of humins via aldol addition and condensation during acid-catalyzed conversion of 5-hydroxymethylfurfural. Energy and Fuels, 2011, 25 (10): 4745-4755.

[18] Hu X, Li C Z. Levulinic esters from the acid-catalysed reactions of sugars and alcohols as part of a bio-refinery. Green Chemistry, 2011, 13 (7): 1676-1679.

[19] Zhang M, Yang H, Liu Y N, Sun X D, Zhang D K, Xue D F. First identification of primary nanoparticles in the aggregation of HMF. Nanoscale Research Letters, 2012, 7 (38): 1-5.

[20] Chuntanapum A, Matsumura Y. Char formation mechanism in supercritical water gasification process: a study of model compounds. Industrial and Engineering Chemistry Research, 2010, 49 (9): 4055-4062.

[21] Yin S D, Pan Y L, Tan Z C. Hydrothermal conversion of cellulose to 5-hydroxymethyl furfural. International Journal of Green Energy, 2011, 8 (2): 234-247.

[22] Srokol Z W, Rothenberg G. Practical issues in catalytic and hydrothermal biomass conversion: concentration effects on reaction pathways. Topics in Catalysis, 2010, 53 (15-18): 1258-1263.

[23] Dee S J, Bell A T. A study of the acid-catalyzed hydrolysis of cellulose dissolved in ionic liquids and the factors influencing the dehydration of glucose and the formation of humins. ChemSusChem. 2011, 4 (8): 1166-1173.

[24] Alonso D M, Wettstein S G, Mellmer M A, Gurbuz E I, Dumesic J A. Integrated conversion of hemicellulose and cellulose from lignocellulosic biomass. Energy and Environmental Science, 2013, 6 (1): 76-80.

[25] Wettstein S G, Alonso D M, Chong Y, Dumesic J A. Production of levulinic acid and γ-valerolactone (GVL) from cellulose using GVL as a solvent in biphasic systems. Energy and Environmental Science, 2012, 5 (8): 8199.

[26] S J, Luterbacher, Rand J M, Alonso D M, Han J, Youngquist J T, Maravelias C T, Pfleger B F, Dumesic J A. Nonenzymatic sugar production from biomass using biomass-derived γ-valerolactone. Science, 2014, 343 (6168): 277-280.

[27] Khazraji A C, Robert S. Self-assembly and intermolecular forces when cellulose and water interact using molecular modeling. Journal of Nanomaterials, 2013, 2013: 1-12.

第四章

含氧化合物水热降解生成焦炭的机理

4.1 引言

在催化水热转化过程中，木质纤维素中的纤维素和半纤维素被水解成单糖，并进一步转化为有价值的平台化学品，如 HMF[1,2]、糠醛、乙酰丙酸[3,4]、二羟基丙酮、甘油醛、丙酮醛和乳酸等[5-7]。遗憾的是，碳水化合物在水热转化过程中会生成一种与褐煤极为相似的固体副产物，即水热焦炭（或者称为胡敏素）[8-12]，大大降低木质纤维素水热处理过程中有价值的平台化学品的收率。水热碳化法（HTC）是一种将木质纤维素转化为固体碳材料的有效方法，所得到的水热焦炭可用于制备新型碳材料，在 CO_2 分离、水净化、储能、催化等领域具有潜在的应用前景[13-18]。因此，研究水热焦炭的结构及其生成机理，对木质纤维素的水热催化炼制过程和水热碳化过程都具有重要意义。

为了揭示焦炭的分子结构，国内外研究人员采用 SEM、热解气相色谱-质谱（PY-GC/MS）、固态^{13}C 核磁共振（solid-state ^{13}C NMR）、傅里叶红外（FT-IR）和元素分析等多种方法分析了由木质纤维素和木质纤维素衍生物（葡萄糖、木糖、蔗糖、淀粉、HMF 和糠醛）生成的焦炭。大量研究表明，虽然水热碳化条件和水热碳化原料会对水热焦炭的结构有一定的影响，但是水热焦炭都是含有多种含氧官能团（如羟基、羧基和羰基等）、以不饱和多环结构为骨架的聚合物[19-24]。关于焦炭的分子结构，目前主要有两种不同的结构模型（图 4-1）。Sevilla 等认为，碳水化合物水热碳化过程中形成的焦炭是以多环芳烃为骨架的、含有大量活性/亲水性氧官能团（即羟基、羰基、羧基和酯基）的聚合物[10,25][图 4-1(a)]。但是，Van Zandvoort 等认为，水热焦炭（在他们的文章中称之为胡敏素）是以呋喃环为骨架的、含有大量含氧官能团的聚合物[19,20][图 4-1(b)]。虽然这两种分子结构模型都被学术界广泛接受，但是，图 4-1(a) 所示分子模型的元素组成与多篇文献所述的焦炭不一致[26-31]，而图 4-1(b) 所示的分子模型缺少苯环片段，这也与多篇文献报道的结果不一致[26,30,32]。因此，这两种焦炭的分子模型都不能令人满意。

根据对水热焦炭结构的表征和对碳水化合物水热降解过程中形成的中间产物的鉴定，人们提出了下述木质纤维素生成水热焦炭的途径[10,18,27,33]：①木质纤维素中的纤维素和半纤维素水解生成葡萄糖、木糖等单糖；②单糖脱水生成 HMF 和糠醛等呋喃类化合物；③HMF 和糠醛等呋喃类化合物经历一系列缩聚反应而形成聚呋喃类化合物；④所形成的聚合物进一步芳构化，形成具有

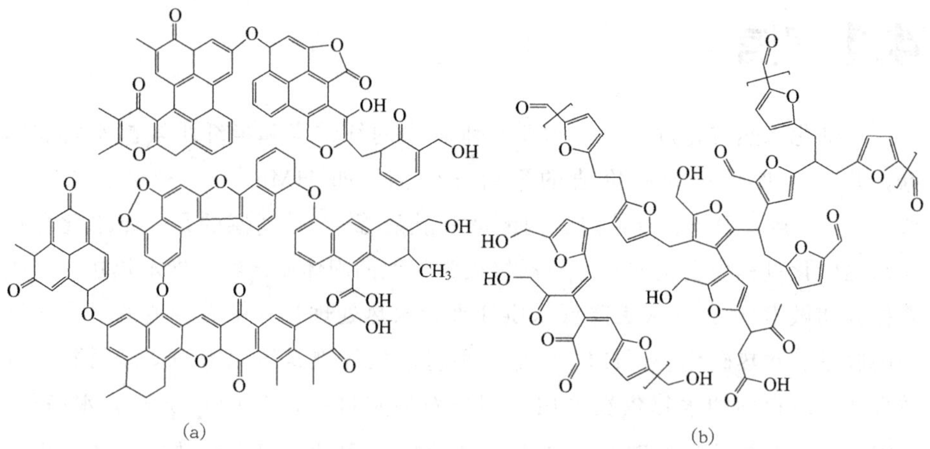

(a)　(b)

图 4-1　文献提出的两种水热焦炭的分子结构[19, 25]

多环芳烃骨架的固态水热焦炭。然而，关于形成水热焦炭的关键前驱体以及前驱体缩聚的初始步骤的细节仍不清楚。Sumerskii 等提出碳水化合物脱水所生成的呋喃类化合物（糠醛和 HMF）分子间的醚化或缩醛反应可能是形成水热焦炭的关键步骤[21]（图 4-2 中路径 1）。Patil 等提出了由糠醛和 HMF 水解开环形成的链式醛，如 2,5-二氧代-6-羟基己醛（DHH），是生成水热焦炭的关键中间体，而这些链式醛的羟醛缩合是碳水化合物缩合成水热焦炭的初始步骤[34,35]（图 4-2 中路径 2）。Dee 等认为，水热焦炭是通过糠醛/HMF 与碳水化合物经缩醛反应而形成的（图 4-2 中路径 3）[36]。Cheng 等提出碳水化合物与糠醛/HMF 的羟醛缩合是水热焦炭形成的关键步骤（图 4-2 中路径 4）[28]。可见，虽然人们对水热焦炭的形成机理进行了大量的研究，但尚未得到明确的结论。

确定碳水化合物降解过程中形成水热焦炭的关键前驱体对揭示水热焦炭的形成机理具有重要意义。为了确定形成水热焦炭的关键前驱体，本章首先研究了多种小分子含氧模型化合物（碳水化合物、呋喃衍生物、环酮、羧酸和含 2～4 个碳原子的短链含氧有机物等）在水热条件下的降解行为并对其水热转化路径进行分析，试图明确碳水化合物生成水热焦炭的关键中间体。研究发现，这些含氧模型化合物在降解过程中，只有能够生成链式 α-羰基醛、α-羰基酸和 β-不饱和醛的模型化合物能够在降解过程生成水热焦炭，而不能够生成链式 α-羰基醛、α-羰基酸和 β-不饱和醛的模型化合物均难以生成水热焦炭，表明碳水化合物在降解过程中生成的 α-羰基醛是碳水化合物生成水热焦炭的关键前驱体（图 4-2 中路径 5）。随后，采用元素分析、FT-IR 和固态 ^{13}C NMR 等

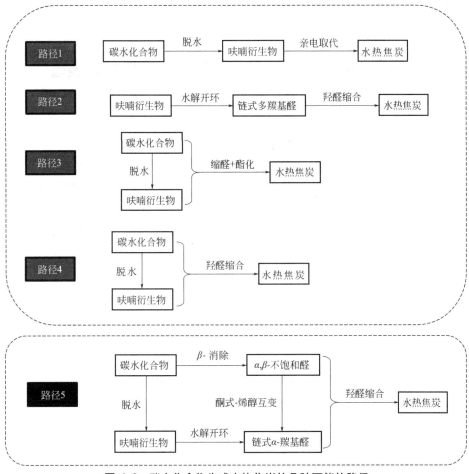

图 4-2 碳水化合物生成水热焦炭的几种可能的路径

几种手段分析了不同原料（葡萄糖、果糖、木糖、核糖、HMF、糠醛、1,3-二羟基丙酮和丙酮醛）在不同溶剂（水和乙酸乙酯）中所得到的焦炭的元素组成和分子结构。元素分析表明，所有水热焦炭的分子式都可以近似表示为 $(C_3H_2O)_n$，而碳水化合物在水中生成的焦炭比在乙酸乙酯中生成的焦炭含有更多的碳和更少的氧。FT-IR 分析和固态 ^{13}C NMR 技术分析结果表明，所有的焦炭都含有多环芳烃结构、酚环结构、呋喃环结构、脂肪 C—O 结构和羰基/羧基。其中，在水中生成的焦炭含有较多的酚环结构，而在乙酸乙酯中生成的焦炭含有较多的呋喃环结构。在对焦炭的结构进行表征的基础上，笔者提出了一种含多环芳烃结构、酚环结构、呋喃环结构和脂肪碳碎片的焦炭分子结构模型。结合水热焦炭的结构分析和对模型化合物的聚合路径分析，笔者提出，水热焦炭是碳水化合物在水热转化过程中所生成的 α-羰基醛通过羟醛缩合、缩醛环化、脱水等过程所生成的。

4.2 模型化合物的水热转化过程及产物的分析表征

4.2.1 模型化合物的水热转化过程

将含有 0.12mol 碳原子的模型化合物和 30mL 溶剂加入微型水热反应釜中，并在 220℃下保持 5h 以进行降解。在转化过程结束后，过滤，用蒸馏水和乙醇洗涤固体产物，并在 105℃下进行干燥称重，确定固体焦炭的质量。滤液和洗涤液则在 105℃下蒸发和干燥，以获得不挥发产物的质量（以水为反应介质时在 130℃进行干燥）。水热焦炭的碳收率是通过将固体残渣中的碳总质量（假设固体残渣中的碳含量为 65%）除以原料中碳的总质量来计算的，而不挥发产物的碳收率则是通过将不挥发产物中的碳总质量（假设不挥发产物中的碳含量应为原料中总碳质量的 60%）除以原料中碳的总质量进行计算，具体计算公式如下：

$$碳收率_{水热焦炭} = \frac{残渣质量 \times 65\%}{1.44} \times 100\%$$

$$碳收率_{不挥发物} = \frac{不挥发物质量 \times 60\%}{1.44} \times 100\%$$

$$碳收率_{可挥发物} = 100\% - 碳收率_{水热焦炭} - 碳收率_{不挥发物}$$

4.2.2 原料的定量分析方法

（1）HPLC 分析

采用 Agilent 1200 系列高效液相色谱仪（Bio-Rad HPX-87H）测定了反应后混合液中糖类和呋喃衍生物的浓度。以 5mmol/L 硫酸水溶液为洗脱液，流速为 0.5mL/min，采用示差折光检测器和紫外检测器（210nm）进行检测。柱温和示差折光检测器温度分别设置为 55℃和 45℃。

（2）GC 分析

采用气相色谱（GC）分析其他研究模型化合物的浓度。GC 分析在 Agilent 7890B 气相色谱仪上进行，色谱柱为 HP-5MS 超惰性毛细管柱（30m×0.25mm×0.25μm）。色谱柱箱温度首先保持在 45℃，随后以 10℃/min 的速率升温到 220℃，并在 220℃保持 2min。载气为氦气，流速为 1.2mL/min。

4.2.3 焦炭的结构表征方法

（1）元素分析

固体焦炭的元素组成在 Vario-EL cube 元素分析仪上（CHNS 模式）进行，采用热导检测器。焦炭中的氧含量通过总质量与检测到的 C、H、N 和 S 的元素之差进行计算。

（2）FT-IR 分析

焦炭的 FT-IR 光谱利用 IR-Prestige21 光谱仪（Shimadzu）分析，并在 $400\sim4000cm^{-1}$ 范围内扫描，分辨率为 $2cm^{-1}$。将样品粉末与 KBr 混合并将其压成薄片，制备 FT-IR 分析样品。

（3）固态^{13}C NMR 分析

采用交叉极化（CP）、魔角旋转（MAS）和大功率^{1}H 去耦技术，在 Agilent 600 DD2 光谱仪（美国 Agilent，磁场强度=14.1T）上记录固体焦炭的固态^{13}C CP/MAS 核磁共振谱。分析过程中，将粉末样品放入 4.0mm 氧化锆转子中，以 10kHz（$4.2\mu s$，$90°$脉冲）、2ms CP 脉冲和 3s 的循环延迟获得光谱，并以 0ppm 下四甲基硅烷（TMS）的 C 信号作为^{13}C 化学位移的校正。

4.3 呋喃衍生物的水热降解行为及降解途径

大量文献认为，碳水化合物在水热条件下降解所生成的呋喃衍生物，如 HMF 和糠醛，在水热焦炭的形成过程中发挥重要作用。为了弄清这些呋喃类化合物生成水热焦炭的路径，笔者首先比较了 HMF、糠醛、5-甲基糠醛、糠醇、糠酸、2-乙酰呋喃、2-甲基呋喃、呋喃这 8 种含有呋喃环结构的物质在以水为反应溶剂时的降解行为（表 4-1）。结果表明，HMF、5-甲基糠醛、糠醛、糠醇和糠酸在相同的水热条件分别能产生 64.6%、12.1%、23.4%、29.2%、11.8%的水热焦炭，而 2-乙酰呋喃、2-甲基呋喃和呋喃在水热条件下没有形成水热焦炭，表明上述几种呋喃类化合物在水热条件下生成焦炭的难易程度有显著的区别。在相同的反应条件下，HMF 和糠醇较易生成水热焦炭，5-甲基糠醛、糠醛和糠酸相对不易生成焦炭，而 2-乙酰呋喃、2-甲基呋喃和呋喃则难以形成水热焦炭。一些研究者提出碳水化合物与呋喃类化合物之间的缩醛/羟醛缩合反应是碳水化合物生成水热焦炭的关键步骤[28,36]，然而，仅由这些呋喃类物质在水热条件下就能形成水热焦炭，表明碳水化合物不是水热焦炭生成所必需的。因此，可以排除碳水化合物与呋喃类物质通过缩醛/羟醛缩合直接生成水热焦炭的可能性。

表 4-1 呋喃衍生物的水热降解行为

降解原料	分子结构	溶剂	转化率/%	碳收率/%		
				水热焦炭	不挥发物	挥发物
HMF		水	100	64.6	14.5	20.9
5-甲基糠醛		水	62.3	12.1	19.2	68.7
糠醛		水	74.6	23.4	12.4	64.2
糠醇		水	100	29.2	30.2	40.6
糠酸		水	—	11.8	14.8	73.4
2-乙酰呋喃		水	—	0	2.7	97.3
2-甲基呋喃		水	—	0	26.5	73.5
呋喃		水	—	0	4.7	95.3
HMF		乙酸乙酯	84.6	0	10.5	89.5
5-甲基糠醛		乙酸乙酯	5.9	0	3.6	96.4
糠醛		乙酸乙酯	4.8	0	1.2	98.8
糠醇		乙酸乙酯	38.3	0	2.9	97.1

一般来说，HMF 和糠醛被认为是碳水化合物生成水热焦炭的关键中间体。国外研究者提出了通过 HMF/糠醛生成水热焦炭的两种可能路径：第一种路径中，这些化合物不发生水解开环反应，直接在这些物质的呋喃环上发生亲电取代反应而聚合成水热焦炭[21,37]（图 4-2 中路径 1）；第二种路径中，HMF/糠醛通过水解开环形成多羰基的链式醛，而这些链式多羰基醛则继续发生缩合反应而生成水热焦炭[34,35]（图 4-2 中路径 2）。为了验证水解开环反应是否为呋喃衍生物生成水热焦炭的必经步骤，笔者进一步比较了 HMF、糠醛、5-甲基糠醛和糠醇这四种呋喃衍生物在水和纯乙酸乙酯中的降解行为。可以看到，这四种呋喃衍生物在乙酸乙酯中降解时的转化率始终低于在水中的转化率，表明这些呋喃衍生物在乙酸乙酯中比在水中稳定。另外，HMF、5-甲基糠醛、糠醛和糠醇在水中转化时都生成了大量的水热焦炭，但是它们在乙酸乙酯中转化时都没有形成水热焦炭，表明这些呋喃衍生物在乙酸乙酯中生成焦炭的路径受到了抑制。这些呋喃衍生物在乙酸乙酯中生成的不挥发物的碳收率（1.3%～10.5%）也比在水中（12%～30%）低得多，表明这些呋喃衍生物（或其降解产物）的缩合/聚合反应被抑制。上述结果证明，水是这些呋喃衍生物形成水热焦炭所必需的。因此，笔者认为，这些呋喃衍生物水解开环生成的链式多羰基化合物是这些呋喃衍生物生成水热焦炭的关键前驱体（图 4-2 中路径 2）。

前面的研究发现，在相同的水热条件下，HMF、糠醛、5-甲基糠醛、糠醇和糠酸可以生成水热焦炭，而 2-乙酰呋喃、甲基呋喃和呋喃难以生成水热焦炭。这可能是因为 HMF、糠醛、5-甲基糠醛、糠醇和糠酸这几种物质易发生水解开环反应，而 2-乙酰呋喃、甲基呋喃和呋喃难以发生水解开环反应，也可能是后面几种物质发生水解开环反应生成的链式含氧化合物不易发生聚合反应。

因为水热焦炭是由某一类具有特殊官能团的物质通过某一系列特殊反应所生成的，所以，分析这些呋喃类化合物的降解途径，并试图找出生成焦炭的物质在降解过程中生成的某类具有特殊官能团的中间体，对于揭示水热焦炭的生成机理具有一定的意义。

图 4-3 为 HMF 的水热降解路径。经过水解开环和酮式-烯醇异构反应后，HMF 可产生两种 α-羰基醛，即 2,5-二氧代-6-羟基-己醛和 2,5-二氧代-3-己烯醛[34,35]。其中，2,5-二氧代-3-己烯醛可通过 C—C 键水解断裂生成甲酸和 4-氧代-2-戊烯醛，而 4-氧代-2-戊烯醛则进一步经过水合重排转化为乙酰丙酸（图 4-3 中路径 1）。此外，由于 HMF 生成的 2,5-二氧代-6-羟基-己醛和 2,5-二

氧代-3-己烯醛中都含有多个活性 α-H，因此它还可以发生分子内羟醛缩合和酮式-烯醇互变异构而生成 1,2,4-苯三酚[38]。同样地，由于两个相邻羰基的强亲电作用，这两种链式 α-羰基醛的 β-C 上的 H 具有较强的活性，导致它们易与其他羰基化合物发生羟醛缩合反应而导致碳链增长[34,35]。因此，笔者推断 2,5-二氧代-6-羟基-己醛和 2,5-二氧代-3-己烯醛这两种 α-羰基醛是 HMF 形成水热焦炭的关键前驱体。

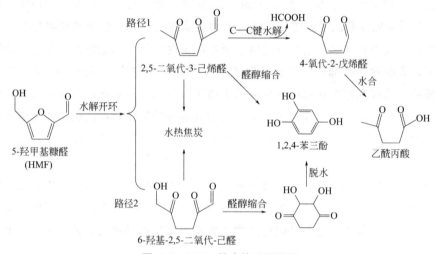

图 4-3　HMF 的水热降解路径

通过与 HMF 降解路径的类比可以得出糠醛、糠醇、糠酸、2-乙酰基呋喃和 2-甲基呋喃的水热降解路径（图 4-4 和图 4-5）。通过水解开环反应，糠醛同样可以得到相应的 α-羰基醛即 2-氧代戊二醛［图 4-4(a)］。与 HMF 水解开环形成的 2,5-二氧-6-羟基-己醛类似，糠醛水解开环形成的 2-氧代戊二醛可以经过 C—C 键水解反应生成甲酸和丁二醛，或者经过羟醛缩合和酮式-烯醇异构化反应，生成环戊酮衍生物。糠醛通过加氢选择性合成环戊酮的方法已经被其他课题组报道过[39,40]。相似地，糠醛水解开环生成的 2-氧代戊二醛这种 α-羰基醛也易与其他羰基化合物发生羟醛缩合反应，所以它也可能是糠醛生成水热焦炭的关键中间体。

糠醇在水解开环后没有生成 α-羰基醛，而是生成了一种 γ-羰基醛即 5-羟基-4-氧代-戊醛，该物质可以通过分子内羟醛缩合和脱水反应而生成戊二酮，或者通过水化重排生成 4,5-二羟基-2-戊烯醛，并进而通过脱水和水合重排而生成乙酰丙酸[41]［图 4-4(b)］。在这些中间产物中，4,5-二羟基-2-戊烯醛可能是糠醇生成焦炭的关键中间体。

糠酸的降解路径相对简单[图 4-4(c)]。糠酸经水解开环反应生成 2,5-二氧代-戊酸，随后 2,5-二氧代-戊酸作为一种 α-羰基酸，可经脱羧生成丁二醛，并进而发生脱水环化转化为呋喃。呋喃在水热降解过程中不会产生水热焦炭（表 4-1 中序列 8），而糠酸的水热降解可以产生水热焦炭，表明 2,5-二氧代-戊酸（一种 α-羰基酸）是糠酸形成水热焦炭的关键中间体。

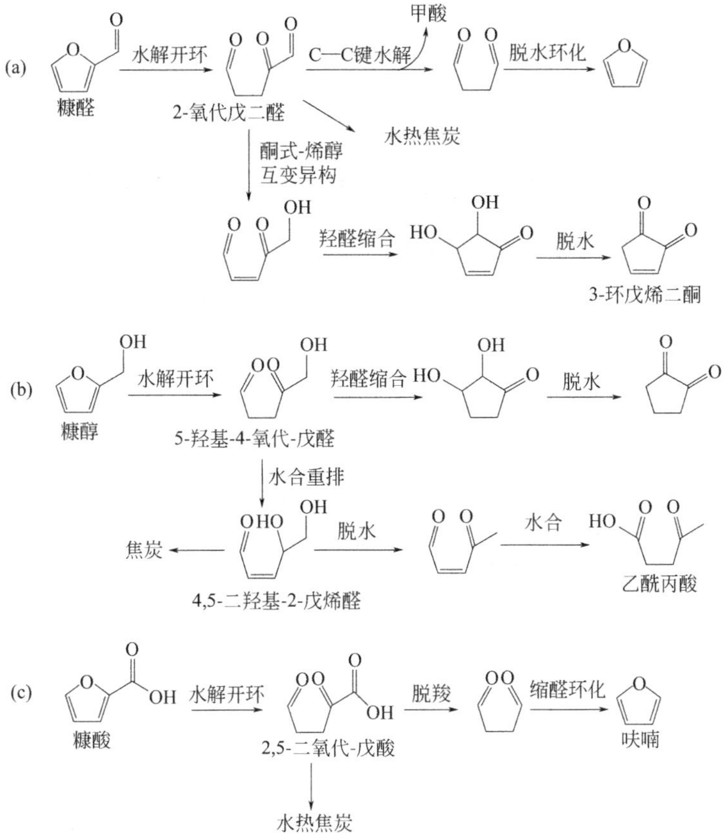

图 4-4　糠醛、糠醇和糠酸的水热降解路径

2-乙酰呋喃和 2-甲基呋喃在水解开环后均不能形成 α-羰基醛和 α-羰基酸（图 4-5）。2-乙酰呋喃水解开环后可以形成一种 α-羰基酮，而 2-甲基呋喃水解开环后可以形成一种 γ-羰基醛。因为 α-羰基酮中的两个羰基分别与甲基和亚甲基（均为给电子基团）相连，而 γ-羰基醛中的酮羰基与醛基没有直接相邻，所以，α-羰基酮和 γ-羰基醛中的 α-H 原子的活性远低于 α-羰基醛和 α-羰基酸中的 α-H 原子，使得 2-乙酰呋喃和 2-甲基呋喃水解开环所生成的链式羰基化合物难以发生羟醛缩合反应。这可能是这两种呋喃衍生在水热条件下难以生成水热焦炭的原因。

图 4-5　2-乙酰呋喃（a）和 2-甲基呋喃（b）的水热降解路径

根据上述对呋喃衍生物水热降解行为的研究，可以总结出下面的规律（图 4-6）：①HMF 和糠醛经水解开环反应后可生成链式 α-羰基醛，而它们在

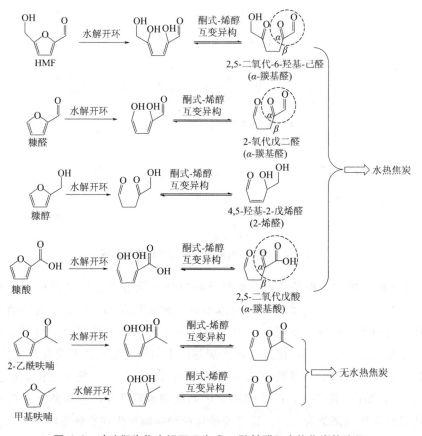

图 4-6　呋喃衍生物水解开环生成 α-羰基醛和水热焦炭的路径

水热降解过程中均易生成水热焦炭；②糠醇在水热降解过程中可生成2-烯醛类物质，它也易生成焦炭；③糠酸经水解开环反应后可生成链式α-羰基酸，而它能够形成少量的水热焦炭；④2-乙酰基呋喃和甲基呋喃在水热条件下经水解开环反应后均不能生成α-羰基醛、α-羰基酸和2-烯醛，而这两种物质在水热转化过程中均不生成水热焦炭。因此，笔者推测，α-羰基醛、2-烯醛和α-羰基酸是水热焦炭形成的关键前驱体。

4.4 碳水化合物的水热降解行为研究及其转化路径分析

上述研究发现呋喃衍生物只能在水中生成水热焦炭，而不能在乙酸乙酯中生成水热焦炭。如果碳水化合物只能通过呋喃衍生物生成水热焦炭，那么碳水化合物在水和乙酸乙酯中的降解行为应该与呋喃衍生物一致。因此，笔者进一步研究了葡萄糖、果糖、山梨糖、木糖和核糖这几种碳水化合物在水和乙酸乙酯中的降解行为。其中，葡萄糖、木糖和核糖属于醛糖，而果糖、山梨糖属于酮糖。

如表4-2所示，这些碳水化合物的水热降解行为与呋喃衍生物有很大不同。所研究的碳水化合物在水和乙酸乙酯中的转化率均达到100%，说明碳水化合物在这两种溶剂中都是不稳定的。此外，这些碳水化合物在水和乙酸乙酯中都能生成大量的水热焦炭，说明这些碳水化合物在乙酸乙酯中生成水热焦炭的反应没有受到抑制。当以水为反应介质时，果糖、山梨糖、葡萄糖、木糖和核糖生成的焦炭收率分别为54.1%、52.8%、60.7%、42.5%和41.0%。当以乙酸乙酯为溶剂时，果糖和山梨糖（均为酮糖）的水热焦炭收率分别降至33.1%和35.8%，而葡萄糖、木糖和核糖（均为醛糖）的水热焦炭收率分别增加到73.1%、56.6%和54.9%，表明酮糖在乙酸乙酯中产生的水热焦炭比在水中产生的水热焦炭少，而醛糖则相反。酮糖和醛糖在水和乙酸乙酯中生成焦炭的区别可以根据这些糖生成呋喃衍生物的难易程度来解释。一般情况下，酮糖在溶剂中以呋喃糖型结构存在，因此它们较易脱水形成呋喃衍生物[42,43]，而呋喃衍生物在乙酸乙酯中不能形成水热焦炭，这就导致酮糖（果糖和山梨糖）在乙酸乙酯中生成的水热焦炭比在水中形成的水热焦炭少。相反，醛糖脱水生成呋喃衍生物的选择性较低，这就导致醛糖在乙酸乙酯中同样生成大量的水热焦炭[2,44]。这些碳水化合物在水和乙酸乙酯中都产生了约10%~20%的不挥发物，表明这些碳水化合物降解生成的中间产物的聚合反应在乙酸乙酯中不能被有效地抑制。

表 4-2 碳水化合的水热降解行为

原料	分子结构	溶剂	转化率/%	碳收率/%		
				水热焦炭	不挥发物	挥发物
果糖		水	100	54.1	16.2	29.7
山梨糖		水	100	52.8	15.2	32
葡萄糖		水	100	60.7	13.5	25.8
木糖		水	100	42.5	12.5	45
核糖		水	100	41.0	13.2	45.8
果糖		乙酸乙酯	100	33.1	12.9	54
山梨糖		乙酸乙酯	100	35.8	12.1	52.1
葡萄糖		乙酸乙酯	100	73.1	12.5	14.4
木糖		乙酸乙酯	100	56.6	19.2	24.2
核糖		乙酸乙酯	100	54.9	16.4	28.7

对比碳水化合物和呋喃衍生物在水和乙酸乙酯中的降解行为可以发现，当以乙酸乙酯为反应介质时，碳水化合物仍能产生大量的水热焦炭而呋喃衍生物却不能生成水热焦炭，说明碳水化合物在乙酸乙酯中可以不通过呋喃衍生物而通过其他中间产物生成水热焦炭。前面的研究表明，α-羰基醛、2-烯醛和α-羰基酸很可能是碳水化合物生成水热焦炭的关键前驱体。因此，笔者认为碳水化合物在乙酸乙酯中转化时可能通过生成其他α-羰基醛、2-烯醛和α-羰基酸而形成水热焦炭。

另一方面，国内外的研究发现，醛糖（葡萄糖、木糖、赤藓糖和甘油醛）在水热降解过程中可生成一系列$C_3 \sim C_6$，如α-羰基醛、α-羟基酸、β,γ-不饱和α-羰基醛和β,γ-不饱和-α羟基酸[6,45-53]。Dusselier等[51,52]发现Sn-β催化葡萄糖水热转化过程中可以生成3-脱氧葡萄糖醛酮、2,5,6-三羟基-3-己烯酸和3-脱氧-β-葡萄糖内酯、4-羟基-2-氧代丁醛和2,4-二羟基丁醛。Elliot等[47]报道了木糖催化降解生成3-脱氧木糖醛酮、2,5-二羟基-3-戊烯酸、2,5-二羟基-4-甲氧基-戊酸和2,5-二羟基-3-戊烯酸。在所有已鉴定的碳水化合物降解产物中，3-脱氧葡萄糖醛酮、3-脱氧木糖醛酮、4-羟基-2-氧代丁醛和丙酮醛都是由醛糖通过β-消除反应和酮式-烯醇互变异构反应而生成的链式α-羰基醛[45,47]。因此，在碳水化合物水热降解过程中，α-羰基醛既可以通过呋喃衍生物的水解开环生成，也可以通过碳水化合物的β-消除反应而生成。

把α-羰基醛当作水热焦炭生成的关键前驱体可以合理解释碳水化合物在水和乙酸乙酯中的降解行为[19,34,35]。如图4-7所示，C_5和C_6碳水化合物都可以通过几种途径生成α-羰基醛：①C_5和C_6的碳水化合物经脱水反应生成呋喃类化合物如HMF和糠醛，然后这些呋喃类化合物再经水解开环反应生成α-羰基醛，如2,5-二氧代-6-羟基己醛和2-氧代-戊二醛[34,35,41]；②C_5和C_6碳水化合物可经β-消除和酮式-烯醇互变异构生成α-羰基醛，如3-脱氧葡萄糖醛酮和3-脱氧木糖醛酮[45,46,49,54,55]；③C_5和C_6碳水化合物可以发生逆羟醛缩合反应生成C_3和C_4碳水化合物，再进一步经β-消除和酮式-烯醇互变异构反应生成丙酮醛等α-羰基醛[6,51]。对于HMF和糠醛，它们只能通过水解开环反应生成α-羰基醛，因此它们只能在水作为反应介质时生成水热焦炭。相反，碳水化合物在所有溶剂中都可以通过β-消除反应生成α-羰基醛，如3-脱氧葡萄糖酮和3-脱氧葡萄糖醛酮，故碳水化合物在水和乙酸乙酯中都能够形成水热焦炭。由于C_5和C_6碳水化合物都能在有机溶剂中产生大量的水热焦炭，因此笔者认为碳水化合物通过β-消除和酮式-烯醇互变异构所形成的α-羰基醛，如 3-脱氧葡萄糖酮和 3-脱氧木糖酮，对碳水化合物生成水热焦炭起着更为关键的作用。

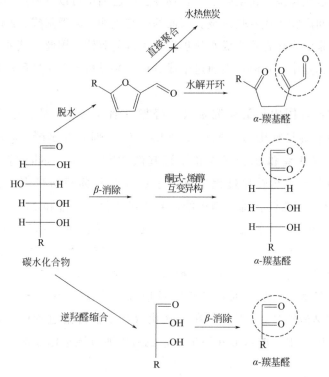

图 4-7　碳水化合物生成 α-羰基醛和水热焦炭的路径（R 表示 H 或 CH_2OH）

4.5　碳水化合物的其他衍生物的水热降解行为

上述研究结果表明，把 α-羰基醛当作生成水热焦炭的关键中间体可以解释呋喃衍生物和碳水化合物在水热转化过程中的行为。碳水化合物在降解过程中还会生成其他物质，如甲酸、乙酰丙酸、乳酸、1,2,4-苯三酚等[5,56,57]，这些化合物也可能是碳水化合物生成水热焦炭的关键中间体。因此，笔者进一步研究了这些模型化合物的水热行为（表 4-3）。

对甲酸、乙酰丙酸和乳酸在水热条件下降解行为的研究发现，这三种羧酸在水热条件下的转化率分别为 19.9％、11.1％和 7.8％，表明这些羧酸的水热稳定性相对较高。这些羧酸在水热降解过中都没有形成固体水热焦炭，说明这些羧酸并不是水热焦炭形成的关键中间体（表 4-3 中序列 1～3）。

表 4-3 碳水化合物的其他衍生物的水热降解行为

反应物	分子结构	转化率/%	碳收率/%		
			水热焦炭	不挥发物	挥发物
乳酸		7.8	0	4.4	95.6
甲酸		19.9	0	0	100
乙酰丙酸		11.1	0	0	100
2-甲基-1,3-环戊二酮		—	0	11.9	88.1
1,2-环戊二酮		—	0	62.8	37.2
1,3-环己二酮		—	0	70.1	29.9
1,2,3-苯三酚		—	0	6.9	93.1
山梨醇		67	0	28.9	71.1
葡萄糖酸		100	17.4	14.7	67.9

有文献认为碳水化合物在水热降解过程中生成的环酮衍生物可能是碳水化合物生成水热焦炭的关键前驱体[58,59]。因此，笔者进一步研究了 2-甲基-1,3-环戊二酮、1,2-环己二酮、1,3-环己二酮和 1,2,3-苯三酚等几种环酮化合物在水热条件下的降解行为，结果也同样列于表 4-3 中。可以看到，这些环酮衍生

物在水热条件下同样没有形成固体水热焦炭，表明这些环酮化合物不是水热焦炭形成的关键前驱体。

葡萄糖加氢可形成山梨醇[60]，葡萄糖氧化可产生葡萄糖酸和葡萄糖酸内酯[61,62]，因此笔者同样研究了山梨醇和葡萄糖酸内酯的水热降解行为。有趣的是，山梨醇没有形成水热焦炭，而葡萄糖酸内酯也仅生成了 17.4% 的水热焦炭。众所周知，山梨醇缺乏醛基，且山梨醇在水热条件下降解时也很难形成醛或酮[63]，这可能是山梨醇在水热转化过程中难以生成水热焦炭的原因。与山梨醇相比，葡萄糖酸在水热条件下可以形成链式多羰基化合物[3]。如图 4-8 所示，葡萄糖酸可以通过 β-消除和酮式-烯醇互变异构形成 3-脱氧葡萄糖酮酸[45,49,54]。3-脱氧葡萄糖酮酸是一种 α-羰基酸，它可以进一步脱羧生成 3,4,5-三羟基-戊醛（或称为 2-脱氧核糖）[64,65]，并进一步转化为糠醇和乙酰丙酸[22]（图 4-8 中路径 1）。另一方面，3-脱氧葡萄糖酮酸也可以通过缩醛环化脱水生成 5-羟甲基糠酸，再通过脱羧反应生成糠醇（图 4-8 中路径 2）。前面的研究已经表明糠醇在水热降解过程中可以生成水热焦炭，而 3-脱氧葡萄糖酮酸是一种 α-羰基酸，所以糠醇和 3-脱氧葡萄糖酮酸应当是葡萄糖酸生成水热焦炭的中间体。

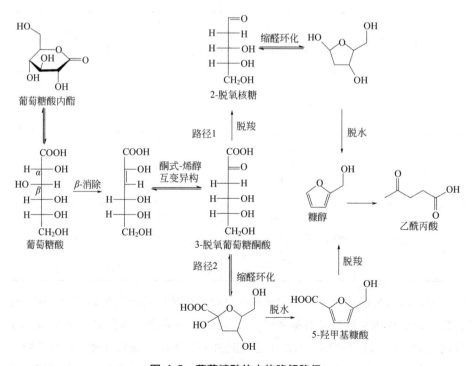

图 4-8　葡萄糖酸的水热降解路径

4.6 $C_2 \sim C_4$ 短链含氧有机物的水热降解行为

短链的含氧有机物的水热降解路径较为简单，因此，研究短链含氧有机物的水热降解行为并对其降解路径进行分析，更能够判断出这些化合物生成水热焦炭的关键前驱体。乙二醇、乙醛、乙二醛和乙醇酸是四种碳链长度为 2 的含氧有机物，1,2-丙二醇、甘油、丙酮、丙醛、丙酮醛、1,3-二羟基丙酮、羟基丙酮和丙酮酸是几种碳链长度为 3 的含氧有机物，而正丁醛、3-羟基-2-丁酮和丁二酮是碳链长度为 4 的含氧有机物。笔者进一步研究了这些模型化合物的水热降解行为。

如表 4-4 所示，在所有这些短链含氧有机物中，只有乙二醛、丙酮醛和 1,3-二羟基丙酮能够分别生成 21.7%、44.2% 和 47.8% 的水热焦炭，而其他含氧有机物都不能产生水热焦炭。三种能够生成水热焦炭的物质中，1,3-二羟基丙酮可以通过脱水反应生成丙酮醛，而乙二醛和丙酮醛均为 α-羰基醛。所以，这三种含氧有机物生成水热焦炭的行为都可以通过将 α-羰基醛作为水热焦炭的前驱体来进行解释。

表 4-4 碳链长度为 2～4 的含氧有机物的水热降解行为

反应物	分子结构	碳收率/%	
		水热焦炭	不挥发物
乙二醇	HO—OH	0	—
乙醛		0	0
乙二醛		21.7	3.8
乙醇酸		0	0
丙酮		0	0
丙醛		0	0
丙酮醛		44.2	6.7

续表

反应物	分子结构	碳收率/%	
		水热焦炭	不挥发物
1,3-二羟基丙酮		47.8	14.2
羟基丙酮		0	8.3
丙酮酸		0	29.1
丙三醇		0	—
1,2-丙二醇		0	—
正丁醛		0	0
3-羟基-2-丁酮		0	2.8
丁二酮		0	25.9

在上述研究的多种短链含氧有机物中，乙醛、丙醛、丙酮、羟丙酮、正丁醛、3-羟基-2-丁酮、丁二酮都是羰基化合物，它们都能在催化剂的作用下发生羟醛缩合反应而实现碳链增长。然而，这些羰基化合物在水热条件下都没有产生水热焦炭。这可能是由于这些羰基化合物发生羟醛缩合反应的活性比 α-羰基醛（酸）低，因此，它们在没有催化剂时在水热条件下难以发生羟醛缩合反应生成固体水热焦炭。需要说明的是，在硫酸的催化作用下，丁二酮可以在水热环境中生成一定量的水热焦炭，表明该物质在催化剂的作用下可以发生聚合反应。但是，在缺少催化剂作用时，该物质难以生成水热焦炭。

上述短链含氧有机物的水热降解行为进一步证实了 α-羰基醛（酸）是碳

水化合物形成水热焦炭的关键前驱体。只有α-羰基醛（酸）或者能够生成α-羰基醛（酸）的物质在水热条件下才易生成水热焦炭，而不能够在水热条件下生成α-羰基醛（酸）的物质在水热条件下则难以生成水热焦炭。

4.7 乙二醛和丙酮醛的水热降解路径分析

乙二醛和丙酮醛是两种最简单的α-羰基醛，它们在水热条件下的降解路径代表了所有α-羰基醛在水热转化过程中会发生的共同反应。因此，笔者分析了这两种化合物的水热转化路径。

如图4-9所示，由于两个相邻羰基的强亲电效应，乙二醛的醛基易被水分子攻击形成一个水合中间体，该中间体可进一步发生1,2-氢转移反应而生成一种α-羟基酸，即乙醇酸（图4-9中路径1）。这就是众所周知的坎尼扎罗反应[66]。事实上，在没有任何催化剂的情况下，乙二醛在水热降解过程中可生成约76%的乙醇酸。另一方面，由于两个羰基的强亲电效应，乙二醛的氢原子具有很强的活性，使得乙二醛易发生羟醛缩合而形成含有大量羰基和羟基的长碳链聚合物，而该长碳链聚合物中的羟基和羰基使得它易发生分子间缩醛环化反应和脱水反应，进而形成以呋喃环为骨架、含有大量羟基和羰基的聚合物（图4-9中路径2）。由于这些聚合物的分子结构信息与文献报道的固体水热焦炭的分子结构信息相一致[19,20]，因此，笔者认为乙二醛在水热过程中生成的水热焦炭涉及乙二醛的羟醛缩合反应、缩醛环化反应和脱水反应等步骤。

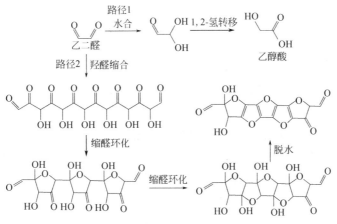

图4-9 乙二醛的水热降解路径

与乙二醛类似，丙酮醛也可以通过坎尼扎罗反应（图4-10中路径1）生成α-羟基酸（乳酸）[6,7,56]，或者通过羟醛缩合、缩醛环化和脱水反应这三个步骤生成以呋喃环为骨架、富含羟基和羰基的聚合物（图4-10中路径2）。丙酮醛在水热降解过程中生成的乳酸收率仅为40%，而生成的水热焦炭收率高达44.2%，这可能是由于丙酮醛分子中的甲基具有给电子效应，降低了其发生坎尼扎罗反应的活性，从而使得更多的丙酮醛分子在水热转化过程中生成水热焦炭。

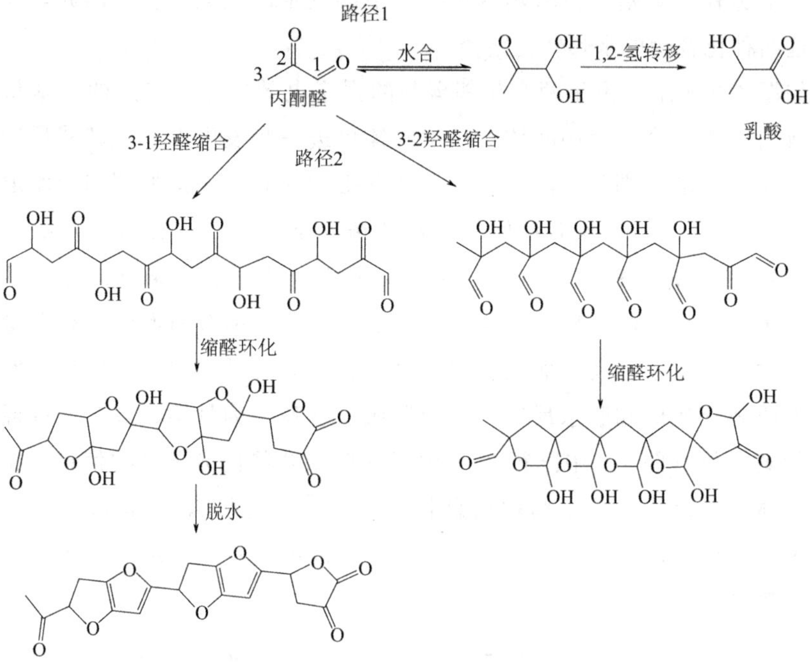

图 4-10 丙酮醛的水热降解路径

对乙二醛和丙酮醛的转化路径的分析表明，在水热条件下，α-羰基醛可以通过两种主要途径被转化：①α-羰基醛可以经水合反应和1,2-氢转移而转化为相应的α-羟基酸，这就是众所周知的坎尼扎罗反应；②α-羰基醛可与其他羰基化合物发生羟醛缩合，随后通过缩醛环化反应和脱水反应而生成固体水热焦炭。另外，如第二章所述，在HMF水热转化生成乙酰丙酸的过程中涉及2,5-二氧代-3-己烯醛（一种α-羰基醛）中的两个羰基之间发生C—C键水解断裂反应[41]，而葡萄糖经3-脱氧葡萄糖醛酮生成HMF时则涉及α-羰基醛发生分子内的缩醛环化反应而生成含有呋喃环的有机物[67]。所以，α-羰基醛还易发生C—C键水解断裂反应和缩醛环化反应。

4.8 水热焦炭的结构分析

4.8.1 水热焦炭的元素分析

表4-5列出了C_6化合物(葡萄糖、果糖、山梨糖、HMF)、C_5化合物(木糖、核糖和糠醛)和C_3化合物(丙酮醛和1,3-二羟基丙酮)分别在水(H_2O)、乙酸乙酯(EAC)和四氢呋喃(THF)中所生成的水热焦炭的元素组成。可以看到,所有水热焦炭的碳含量都约为62.5%~70.2%,氢含量约为2.7%~4.0%,氧含量约为26.0%~33.9%,这与之前的文献报道结果相一致[19,25-29]。在乙酸乙酯和四氢呋喃中形成的焦炭的元素组成非常相似,说明这些有机溶剂都是惰性的,不参与焦炭的形成过程。但是,在水中形成的焦炭比在有机溶剂中形成的焦炭含有更少的氧和更多的碳,这表明在水中形成的焦炭中氧原子的脱除更彻底。

表4-5 碳水化合物及其衍生物生成的水热焦炭收率及其元素组成

水热焦炭	原料	溶剂	碳收率/%	元素组成/%			原子比例		假想分子式
				C	H	O	H/O	C/O	
Char-PRV-WT	丙酮醛	H_2O	44.2	70.2	3.9	26.0	2.4	3.6	$(C_3H_2O_{0.8})_n$
Char-DHA-WT	1,3-二羟基丙酮	H_2O	47.8	68.9	3.7	27.4	2.1	3.3	$(C_3H_{1.9}O_{0.9})_n$
Char-HMF-WT	HMF	H_2O	64.6	66.0	3.3	30.7	1.7	2.9	$(C_3H_{1.8}O_1)_n$
Char-FF-WT	糠醛	H_2O	23.4	67.5	3.2	29.4	1.7	3.1	$(C_3H_{1.7}O_1)_n$
Char-GLU-WT	葡萄糖	H_2O	60.7	66.1	2.7	31.2	1.4	2.8	$(C_3H_{1.5}O_{1.1})_n$
Char-FRU-WT	果糖	H_2O	54.1	66.5	3.5	30	1.9	3.0	$(C_3H_{1.9}O_1)_n$
Char-XYL-WT	木糖	H_2O	46.4	66.5	3.2	30.3	1.7	2.9	$(C_3H_{1.7}O_1)_n$
Char-RIB-WT	核糖	H_2O	41.0	65.8	3.2	31.1	1.6	2.8	$(C_3H_{1.7}O_{1.1})_n$
Char-HMF-EAC	HMF	EAC	0						
Char-FF-EAC	糠醛	EAC	0						
Char-GLU-EAC	葡萄糖	EAC	73.1	63.2	4	32.8	1.9	2.6	$(C_3H_{2.3}O_{1.2})_n$
Char-FRU-EAC	果糖	EAC	33.1	64.5	3.6	31.9	1.8	2.7	$(C_3H_2O_{1.1})_n$
Char-XYL-EAC	木糖	EAC	56.6	64.3	4	31.8	2	2.7	$(C_3H_{2.2}O_{1.1})_n$
Char-RIB-EAC	核糖	EAC	54.9	64.6	3.9	31.5	2	2.7	$(C_3H_{2.2}O_{1.1})_n$
Char-HMF-THF	HMF	THF	0						
Char-FF-THF	糠醛	THF	0						

续表

水热焦炭	原料	溶剂	碳收率/%	元素组成/%			原子比例		假想分子式
				C	H	O	H/O	C/O	
Char-GLU-THF	葡萄糖	THF	52.6	65.2	3.9	30.9	2.0	2.8	$(C_3H_{2.1}O_{1.1})_n$
Char-FRU-THF	果糖	THF	27.4	64.4	3.8	31.7	1.9	2.7	$(C_3H_{2.1}O_{1.1})_n$
Char-XYL-THF	木糖	THF	38.7	62.5	3.6	33.9	1.7	2.5	$(C_3H_{2.1}O_{1.2})_n$
Char-RIB-THF	核糖	THF	40.4	64.6	3.9	31.4	2.0	2.7	$(C_3H_{2.2}O_{1.1})_n$

根据焦炭的元素组成，可以计算出焦炭中 H/O 和 C/O 的原子比[19,26,31]。如表 4-5 所示，所有焦炭的 H/O 原子比在 1.4~2.4 之间，而 C/O 原子比在 2.6~3.6 之间，表明焦炭均含有约 $3n$ mol 碳原子、$2n$ mol 氢原子和 n mol 氧原子。考虑到每个葡萄糖、果糖、HMF 和 1,2,4-苯三醇均含有 6 个碳原子，而 1,3-二羟基丙酮和丙酮醛都含有 3 个碳原子，笔者认为所有水热焦炭的分子式都可以近似地表示为 $(C_3H_2O)_n$。碳水化合物及其脱水衍生物在水中形成的焦炭的分子式可以表示为 $(C_3H_{2-(0.1\sim0.5)}O_{1\sim1.1})_n$，而这些化合物在惰性有机溶剂（乙酸乙酯和四氢呋喃）中所形成的焦炭的分子式可以近似表示为 $[C_3(H_2O)_{1.1\sim1.2}]_n$。很明显，$C_5$ 和 C_6 化合物在有机溶剂中形成的 H/O 原子比接近 2∶1，而它们在水中形成的 H/O 原子比均低于 2∶1。本研究中采用的碳水化合物原料中的 H/O 原子比均为 2∶1，而脱水、水合、缩醛、醚化、酯化、缩醛缩合等反应均不能改变碳水化合物产物中的 H/O 原子比。相反，涉及 C—C 键断裂的反应可能会改变 H/O 原子比。因此，笔者认为碳水化合物及其衍生物在水溶液中生成水热焦炭时，必定涉及了一些 C—C 键的断裂反应，并且一定生成了一些 H/O 原子比高于 2∶1 的可溶性物质。相反，在有机溶剂中生成的焦炭的 H/O 原子比接近 2∶1，表明在这些有机溶剂中，C—C 键的断裂没有大规模地发生。这也意味着，水分子可能对 C—C 键断裂起到重要作用。

如前所述，研究人员提出了两种不同的焦炭分子结构模型[19,25]。图 4-1(a) 中焦炭分子结构模型的分子式可以表示为 $(C_3H_{2.25}O_{0.58})_n$，它比这里形成的水热焦炭含有更多的 H 和更少的 O，这表明图 4-1(a) 中的水热焦炭分子结构模型与此处形成的焦炭相比过于芳构化。相比之下，图 4-1(b) 所示结构模型的分子式可以表示为 $(C_3H_{2.49}O_{1.11})_n$，更接近于此处形成的水热焦炭的元素组成，表明这种结构模型可能与此处得到的水热焦炭结构更接近。不过，这两种模型的 H/O 原子比都大于 2∶1，表明实际上的水热焦炭比这些水热焦炭分子结构模型含有更少的 H 和更多的不饱和键。

4.8.2 水热焦炭的 FT-IR 分析

各种水热焦炭的 FT-IR 谱图如图 4-11 所示。它们的 FT-IR 谱图非常相似，且与之前的文献报道一致[2,19,25,34,68]。在 3400cm^{-1} 处的宽峰归属于 O—H 拉伸振动，表明所有水热焦炭都含有相当多的羟基[2]。在 2991cm^{-1} 处的吸收峰（归属于脂肪族 C—H 拉伸振动）及 1483cm^{-1} 和 1392cm^{-1}（均归属于脂肪族 C—H 变形振动）处的吸收峰表明所有这些水热焦炭都含有脂肪族碳原子[25]。位于 1633cm^{-1} 的吸收峰（归属于 C=O 的拉伸振动）表明在这些水热焦炭中存在羰基或羧基[2]，而在 1176cm^{-1} 处出现的吸收峰（归属于 O=C—O 基团中的 C—O 拉伸振动）证实了这些水热焦炭中存在羧基或酯基[68]。位于 1598cm^{-1}（归属于芳香族/呋喃环的 C=C 拉伸）、1070cm^{-1} 和 1023cm^{-1}（两者都归属于芳香族或 α-不饱和 C—O 拉伸）及 787cm^{-1}（归属于芳香族面外 C—H 变形）处的峰表明，在水热焦炭中存在酚环/呋喃环结构[19,25]。

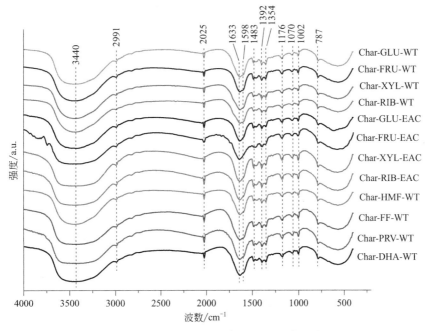

图 4-11 各种水热焦炭的 FT-IR 谱图

对水热焦炭在 700~800cm^{-1} 峰的归属有不同的解释。Van Zandvoort 等认为 700~800cm^{-1} 左右的吸收峰是水热焦炭中存在呋喃环结构的证据[19]，而 Sevilla 等提出 700~800cm^{-1} 左右的峰是水热焦炭中酚存在环结构的证据[25]。然而，呋喃环和酚环结构都能在该区域出现吸收峰，因此，仅依据 700~800cm^{-1} 左右的吸收峰不能确定水热焦炭中是否存在酚环或呋喃环结构。

4.8.3 水热焦炭的固态 ^{13}C NMR 分析

水热焦炭的固态 ^{13}C NMR 谱图如图 4-12 所示。根据文献报道，固态 ^{13}C NMR 波谱可分为四个不同的区域：δ＝0~60（对应于脂肪族 C—C 键的 sp^3 碳原子）、δ＝60~100（对应于脂肪族 C—O 基团的 sp^3 碳原子）、δ＝100~160（对应于芳香环中的 sp^2 碳原子）和 δ＝160~220（对应于羰基/羧基 C＝O 中的 sp^2 碳原子）。如图 4-12 所示，所有这些水热焦炭在 0~60、100~160 和 160~220 的

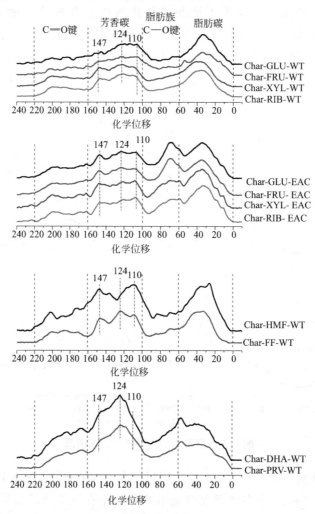

图 4-12 水热焦炭的固态 ^{13}C NMR 谱图

区域内都存在吸收峰，表明所有焦炭都含有饱和脂肪族片段、芳香族结构和羰基/羧基官能团。

固态^{13}C NMR 谱图中，在 $\delta=0\sim60$ 的区域内，由葡萄糖、果糖、木糖、核糖、HMF 和糠醛在水和乙酸乙酯中形成的焦炭均在 $\delta=33\sim35$ 处显示出强烈的峰，这表明这些水热焦炭含有大量的仲碳原子[19]。相比之下，由丙酮醛和 1,3-二羟基丙酮形成的水热焦炭都在 $\delta=42$ 和 $\delta=58$ 处出现明显的峰，但在 $\delta=0\sim40$ 范围内仅出现微弱的峰，表明这些水热焦炭主要含有叔碳和季碳原子[19]。应注意，丙酮醛中含有一个甲基，如果丙酮醛中的甲基不参与水热焦炭的形成过程，则丙酮醛形成的水热焦炭中将会含有大量的甲基碳原子。但丙酮醛形成的水热焦炭中只检测到少量的甲基碳原子，说明丙酮醛的甲基碳原子参与了水热焦炭的形成过程。这一结果再次表明，丙酮醛生成水热焦炭的过程中应当涉及了羟醛缩合反应这一步骤。

固态^{13}C NMR 谱图中 $\delta=60\sim100$ 范围内的峰归属于醇或醚中的脂肪族 C—O 键。可以看到，以水为溶剂形成的水热焦炭和以乙酸乙酯为溶剂所形成的水热焦炭的谱图在这一区域存在明显差异。在乙酸乙酯中形成的水热焦炭的固态^{13}C NMR 谱图在该区域均显示出明显的峰，表明这些水热焦炭中存在大量的脂肪族 C—O 键。相比之下，以水为溶剂形成的水热焦炭在该区域都没有峰，这表明这些水热焦炭缺乏脂肪族 C—O 官能团。注意到上面的 FT-IR 分析已经证实了所有的水热焦炭中都存在大量的羟基，而此处固态^{13}C NMR 谱图表明在水中形成的水热焦炭都缺少脂肪族 C—O 键，表明在水中生成的水热焦炭的羟基应当与芳香结构相连。因此，在水中形成的水热焦炭中应该存在大量的酚环结构。

所有水热焦炭的固体^{13}C NMR 谱图中在 $\delta=100\sim160$ 区域内都出现了峰，表明所有这些水热焦炭产物都具有芳香环结构。一个明显的区别是，丙酮醛和 1,3-二羟基丙酮形成的水热焦炭在 $\delta=124$ 处都显示出一个显著的峰，而 C_5 和 C_6 模型化合物形成的焦炭在 $\delta=147$ 和 $\delta=110$ 处都显示出显著的峰。文献指出，在 $\delta=147$ 和 $\delta=110$ 左右的两个主峰分别与酚环/呋喃环中的 O—$\underline{C}$=CH 和 O—C=$\underline{C}$H 碳有关[26,27,32]，而 $\delta=124$ 左右的特征峰由苯系碳原子产生[20,26,30,32]。因此，由丙酮醛和 1,3-二羟基丙酮形成的水热焦炭应当含有大量的稠环芳烃结构，而由 C_5 和 C_6 化合物形成的水热焦炭含有较多的呋喃环/酚环结构。C_5 和 C_6 化合物所生成的水热焦炭在 $\delta=124$ 处出现了微弱吸收峰，表明这些水热焦炭也都含有少量的稠环芳烃结构。此外，在水中形成的水热焦炭在 $\delta=124$ 处的峰比在乙酸乙酯中形成的水热焦炭更为明显，表明在水中生成的水热焦炭含有更多的稠环芳烃结构。由于多环芳烃碎片的含量是表征芳构化程度的指标，笔者认为丙酮醛和1,3-二羟基丙酮形成的水热焦炭比C_5和C_6形成的水热焦炭的芳构化程度高，而碳水化合物在乙酸乙酯中形成的水热焦炭的芳构化程度最低。

总之，FT-IR分析和固态^{13}C NMR分析均证实，所有水热焦炭都含有稠环芳烃结构、酚环结构、呋喃环结构和饱和脂肪族碎片以及少量的羰基/羧基。在水溶液中形成的水热焦炭富含酚环结构却缺乏脂肪族C—O官能团，而在乙酸乙酯溶液中形成的水热焦炭富含呋喃环结构和脂肪族C—O官能团。多环芳烃、酚类和呋喃类碎片的存在表明，在水热焦炭的形成过程中，一定会发生碳链增长反应。

4.9 水热焦炭的形成机理

前面的研究表明链式α-羰基醛是形成水热焦炭的关键前驱体，而羟醛缩合反应是链式α-羰基醛生成水热焦炭的关键步骤[34,35]。因为丙酮醛、3-脱氧葡萄糖醛酮和2,5-二氧代-6-羟基己醛是葡萄糖降解过程中容易生成的三种α-羰基醛[34,35,41,45-47]，所以笔者以羟醛缩合为起始步骤分析了这三种α-羰基醛的聚合路径。

丙酮醛是葡萄糖降解所形成的最简单的α-羰基醛。如图4-13所示，丙酮醛在C_1和C_2位点上含有两个羰基，导致它的C_3原子上具有活性较高的α-H。因此它可以通过3-1羟醛缩合和3-2羟醛缩合生成两种长链聚合物。由3-1羟醛缩合形成的长链聚合物可通过缩醛环化和脱水反应而生成富含呋喃结构的多环聚合物，或通过1,2-氢转移反应、分子内羟醛缩合反应而生成富含酚环结构的多环聚合物，这两种多环聚合物的分子式都可以表示为$(C_3H_2O)_n$，这与丙酮醛和1,3-二羟基丙酮形成的水热焦炭的元素组成一致。这些聚合物富含芳香碳原子却缺乏脂肪族C—O键，也符合丙酮醛和1,3-二羟基丙酮形成的水热焦炭的固态^{13}C NMR分析结果。这表明，上述丙酮醛经3-1羟醛缩合、缩醛环化和脱水生成聚合物的路径可能是丙酮醛生成水热焦炭的主要路径。相反，丙酮醛经3-2羟醛缩合所生成的长链聚合物只能通过缩醛环化反应而生成缺少芳香族碳原子却富含脂肪族C—O键的多环聚合物$(C_3H_4O_2)_n$，这种聚合物的结构与丙酮醛和1,3-二羟基丙酮形成的水热焦炭的元素组成和固态^{13}C NMR分析不符。根据上述分析，笔者认为丙酮醛分子之间的3-1羟醛缩合是丙酮醛生成水热焦炭的主要途径，而丙酮醛分子之间的3-2羟醛缩合对丙酮醛生成水热焦炭的形成起着不太重要的作用。

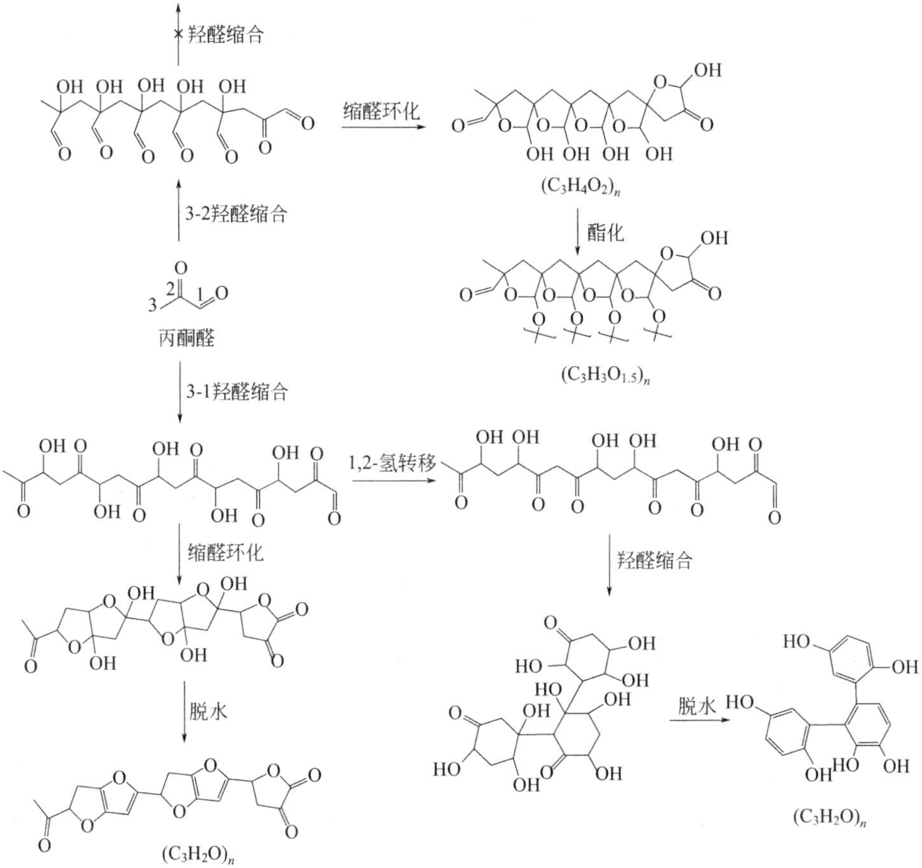

图 4-13 丙酮醛生成水热焦炭的可能路径

图 4-14 展示了 3-脱氧葡萄糖醛酮生成水热焦炭的可能路径。与丙酮醛类似，3-脱氧葡萄糖醛酮也可通过 3-1 羟醛缩合和 3-2 羟醛缩合形成长链聚合物。这些长链聚合物可再经缩醛环化和脱水形成富含呋喃片段和脂肪族 C—O 官能团中的多环聚合物 $(C_6H_6O_3)_n$。随后，这些多环聚合物 $(C_6H_6O_3)_n$ 都可以进一步通过醚化反应生成含有呋喃环结构和脂肪族 C—O 键的聚合物 $(C_3H_2O)_n$，或通过水解开环、1,2-氢转移、分子内羟醛缩合和脱水反应形成富含酚环结构、缺乏脂肪族 C—O 基团的多环聚合物 $(C_3H_2O)_n$。在 3-脱氧葡萄糖醛酮形成富含酚环结构的聚合物的过程中涉及呋喃环聚合物的水解开环反应，这可以解释为什么在水中形成的水热焦炭含有较多的酚环结构，而在乙酸乙酯形成的水热焦炭富含呋喃环结构。此外，对比图 4-13 和图 4-14 所示路径可以发现，由 3-脱氧葡萄糖醛酮生成酚类聚合物的过程比从丙酮醛生成酚

类聚合物的过程涉及更多步骤，这可能是由丙酮醛和1,3-二羟基丙酮形成的水热焦炭比由 C_5 和 C_6 碳水化合物形成的水热焦炭具有更高芳构化程度的原因。

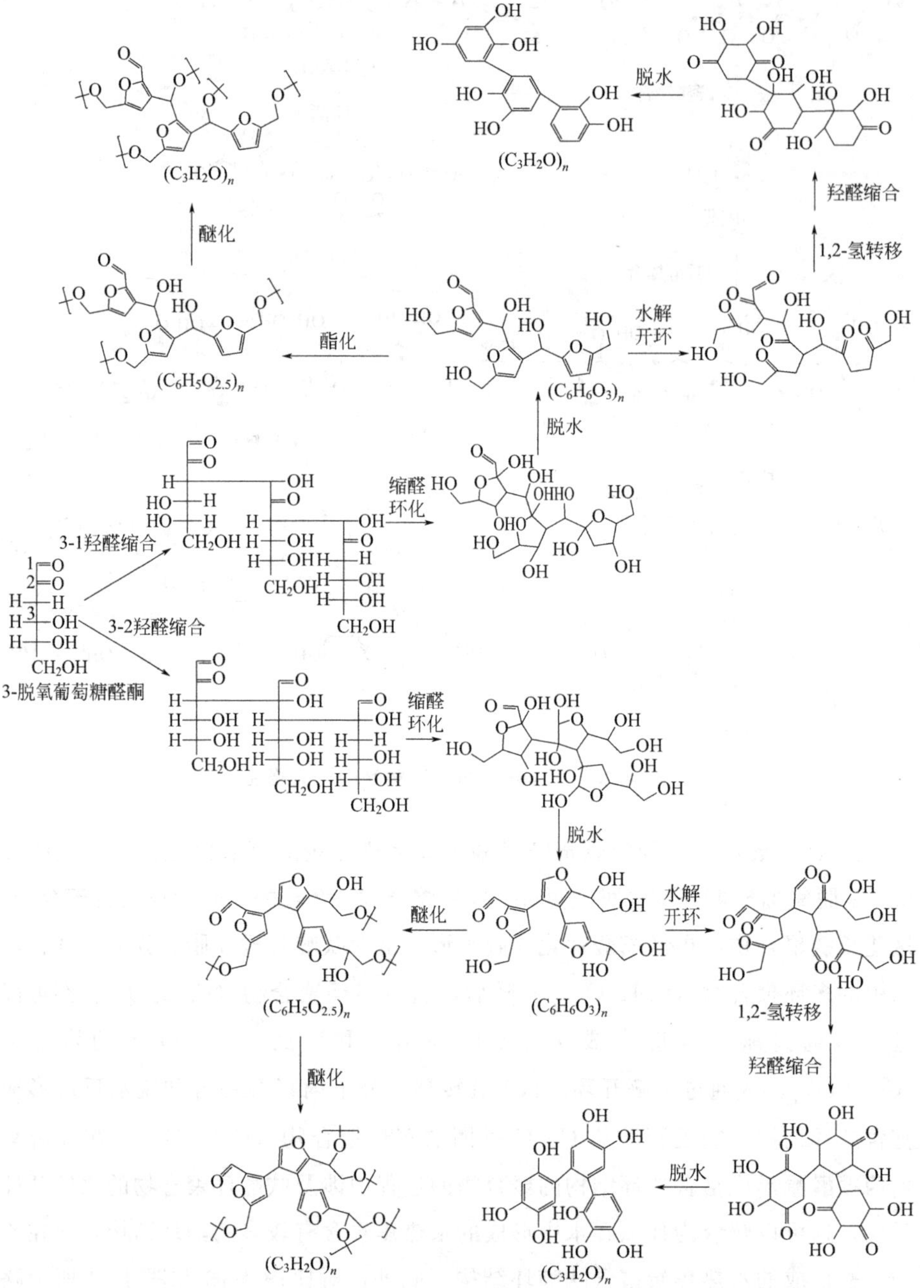

图 4-14　3-脱氧葡萄糖醛酮生成水热焦炭的可能路径

与丙酮醛和 3-脱氧葡萄糖醛酮不同，2,5-二氧代-6-羟基-己醛在 C_1、C_2 和 C_5 原子上均有羰基。因此，2,5-二氧代-6-羟基-己醛可经历 3-1 羟醛缩合、3-2 羟醛缩合、3-5 羟醛缩合三种缩合路径而生成富含羟基的长链聚合物（图 4-15）。类似地，这些长链聚合物也都可以通过缩醛环化反应而生成富含呋喃环结构和脂肪族 C—O 基团的多环聚合物 $(C_6H_6O_3)_n$，而该类多环聚合物都可以进一步通过醚化反应生成富含呋喃环和脂肪族 C—O 结构的多环聚合物 $(C_3H_2O)_n$，或通过水解开环、1,2-氢转移、分子内羟醛缩合和脱水反应而生成富含酚环结构、缺少脂肪族 C—O 键的多环聚合物 $(C_3H_2O)_n$。这两种结构分别与在乙酸乙酯中生成的水热焦炭和在水中生成的水热焦炭结构相一致。

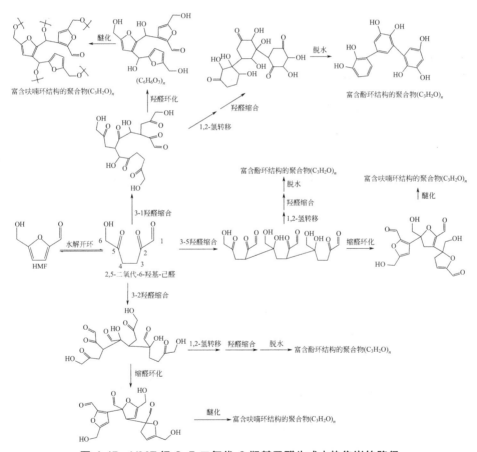

图 4-15　HMF 经 2,5-二氧代-6-羟基己醛生成水热焦炭的路径

通过上面对丙酮醛、3-脱氧葡萄糖醛酮和 2,5-二氧代-6-羟基己醛的聚合路径分析可以看出，在这些 α-羰基醛发生羟醛缩合反应后，都可以生成富含羟基和羰基的链式聚合物，而这些链式聚合物都可以通过缩醛环化和脱水反应而生成富含呋喃环结构的聚合物，或者通过分子内羟醛缩合和脱水反应而生成富含酚环结构的聚合物。由于水热焦炭中同时存在呋喃环结构和酚环结构，因此上述两种路径都是生成焦炭的路径。在以水为溶剂时，聚合物中的呋喃环结构可能发生水解开环、分子内羟醛缩合和脱水反应而转化为酚环结构，因此在以水为溶剂时所生成的水热焦炭含有更多的酚环结构。碱可以催化羟醛缩合反应，所以在碱性水溶液中生成的水热焦炭含有更多的酚环结构[20]。

4.10 水热焦炭的分子结构

根据水热焦炭的特性和形成机理，笔者提出了一个水热焦炭的分子结构模型如图 4-16 所示。该分子结构模型的化学式可以表示为 $(C_3H_2O)_n$，它符合前面测定出的水热焦炭的元素组成；该水热焦炭分子结构模型中含有多环芳烃结构、酚环结构、呋喃环结构和脂肪碳，也符合 FT-IR 和固态 ^{13}C NMR 分析[21,23]。因此，笔者认为该水热焦炭的分子结构是合理的。

需要说明的是，水热焦炭的分子结构并不是固定的，而是随原料的种类、碳化过程中反应条件的苛刻程度而变化[19,25,26,32]。在温和的水热碳化条件下，呋喃环结构和脂肪碳占优势[19]，在苛刻的水热碳化条件下，多环芳烃和酚环结构占优势[25,32]。而在高温热解处理后，呋喃环结构和酚环结构都会转化为多环芳烃结构[26]。

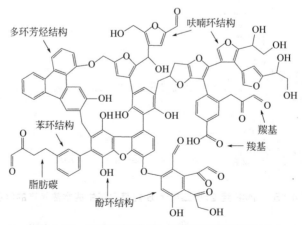

图 4-16　笔者提出的水热焦炭的分子结构模型

4.11 抑制碳水化合物水热降解过程中水热焦炭的形成

数十年来，人们一直在追求将丰富廉价的木质纤维素催化转化为高附加值的平台化学品，如 HMF、乙酰丙酸和乳酸等，但由于碳水化合物在水热过程中易生成水热焦炭，导致上述高附加值的平台化学品的选择性较低，大大降低了木质纤维素水热催化炼制技术的经济性[2,7]。笔者的研究表明，在碳水化合物的水热降解过程中形成的链式 α-羰基醛是水热焦炭的关键前驱体。因此，在木质纤维素水热炼制过程中，抑制 α-羰基醛的生成、降低反应过程中 α-羰基醛的浓度，是抑制水热焦炭生成的有效途径。众所周知，在由水和有机溶剂组成的两相体系中转化碳水化合物时，由于有机相可以迅速地萃取生成的呋喃类物质，从而抑制呋喃类物质发生水解开环反应，能够有效地抑制水热焦炭的生成；在加氢和氧化条件下同样能抑制 α-羰基醛的形成，所以，通过在加氢[69-71]或者氧化[3,72]条件下去转化木质纤维素可以抑制水热焦炭的生成、提高碳原子的利用率。Onda 等[72]在 80℃的 NaOH 水溶液中，采用流动的空气作为氧化剂，用 Pt/Al_2O_3 或 Pt/MgO 催化葡萄糖转化可得 57% 的乳酸并联产 20%～40% 的葡萄糖酸。该体系中，NaOH 的作用是催化葡萄糖经异构化、逆羟醛缩合、β-消除、坎尼扎罗反应而生成乳酸，而 Pt/Al_2O_3 和 Pt/MgO 的作用是催化葡萄糖发生氧化反应而生成葡萄糖酸，从而抑制固体水热焦炭的产生。这一研究与 Lin 等[3]在氧化环境中催化纤维素轻度氧化转化生成乙酰丙酸的思路类似。这两个在水热氧化条件下转化碳水化合物的研究都发现，水热氧化可以抑制水热焦炭的产生。

4.12 结论

通过对多种木质纤维素衍生含氧化合物的水热转化行为的研究和对这些模型化合物的转化途径的分析，笔者证明了碳水化合物水热转化过程中所生成的链式 α-羰基醛是形成水热焦炭的关键中间体。通过对葡萄糖降解途径的分析，发现在葡萄糖的水热转化过程中，可以形成多种 α-羰基醛，如 3-脱氧葡萄糖醛酮、5,6-二羟基-2-氧代-3-己烯醛、2,5-二氧代-6-羟基-己醛和丙酮醛，这些都是葡萄糖水热转化过程中形成水热焦炭的关键前驱体。

利用元素分析、FT-IR 分析和固体 ^{13}C NMR 分析手段对碳水化合物及其脱水产物在水和惰性有机溶剂中形成的水热焦炭的结构进行了表征。元素分析证实,这些水热焦炭的分子结构都可以近似地表示为 $(C_3H_2O)_n$,而 FT-IR 分析和固态 ^{13}C NMR 分析证实所有水热焦炭都有多环芳烃结构、酚环结构、呋喃环结构、脂肪碳和少量羰基或羧基。相比于在有机溶剂中形成的水热焦炭而言,元素分析表明在水中形成的水热焦炭含有较少的氧元素和较多的碳元素,而 FT-IR 和固态 ^{13}C NMR 分析证明在水中形成的水热焦炭含有更多的碳环结构和更少的呋喃环结构。

以 α-羰基醛的羟醛缩合反应作为形成焦炭的初始步骤对丙酮醛、3-脱氧葡萄糖醛酮和 2,5-二氧代-6-羟基己醛的聚合路径进行了分析,并发现 α-羰基醛可以通过羟醛缩合、缩醛环化和脱水反应生成富含呋喃环和脂肪族 C—O 基团的聚合物 $(C_3H_2O)_n$,而这些聚合物中的呋喃环结构可以发生水解开环、分子内羟醛缩合、缩醛环化和脱水反应生成富含碳环结构的聚合物 $(C_3H_2O)_n$。羟醛缩合、缩醛环化、脱水是碳水化合物生成水热焦炭的关键反应步骤。

参 考 文 献

[1] Wang T F, Nolte M W, Shanks B H. Catalytic dehydration of C-6 carbohydrates for the production of hydroxymethylfurfural (HMF) as a versatile platform chemical. Green Chemistry, 2014, 16 (2): 548-572.

[2] Shi N, Liu Q Y, Zhang Q, Wang T J, Ma L L. High yield production of 5-hydroxymethylfurfural from cellulose by high concentration of sulfates in biphasic system. Green Chemistry, 2013, 15 (7): 1967-1974.

[3] Lin H F, Strull J, Liu Y, Karmiol Z, Plank K, Miller G, Guo Z H, Yang L S. High yield production of levulinic acid by catalytic partial oxidation of cellulose in aqueous media. Energy and Environmental Science, 2012, 5 (12): 9773-9777.

[4] Kang S M, Fu J X, Zhang G. From lignocellulosic biomass to levulinic acid: A review on acid-catalyzed hydrolysis. Renewable and Sustainable Energy Reviews, 2018, 94: 340-362.

[5] Wang F F, Liu C L, Dong W S. Highly efficient production of lactic acid from cellulose using lanthanide triflate catalysts. Green Chemistry, 2013, 15 (8): 2091-2095.

[6] Holm M S, Saravanamurugan S, Taarning E. Conversion of sugars to lactic acid derivatives using heterogeneous zeotype catalysts. Science, 2010, 328 (5978): 602-605.

[7] Shi N, Liu Q Y, He X, Cen H, Ju R M, Zhang Y L, Ma L L. Production of lactic acid from cellulose catalyzed by easily prepared solid $Al_2(WO_4)_3$. Bioresource Technology Reports, 2019, 5 66-73.

[8] Zandvoort I V, Ec E R H V, Peinder P D, Heeres H J, Bruijnincx P C A, Weckhuysen B M. Full, reactive solubilization of humin byproducts by alkaline treatment and characterization of the alkali-treated humins formed. Acs Sustainable Chemistry and Engineering, 2015, 3 (3): 533-543.

[9] Maruani V, Stacy Narayanin-Richenapin, Framery E, Andrioletti B. Acidic hydrothermal dehy-

dration of D-Glucose into humins: identification and characterization of intermediates. Acs Sustainable Chemistry and Engineering. 2018, 6 (10): 13487-13493.

[10] Sevilla M, Fuertes A B. Chemical and structural properties of carbonaceous products obtained by hydrothermal carbonization of saccharides. Chemistry—A European Journal. 2009, 15 (16): 4195-4203.

[11] Funke A, Ziegler F. Hydrothermal carbonization of biomass: a summary and discussion of chemical mechanisms for process engineering. Biofuels Bioproducts and Biorefining-Biofpr, 2010, 4 (2): 160-177.

[12] Filiciotto L, Balu A M, Van Der Waal J C, Luque R. Catalytic insights into the production of biomass-derived side products methyl levulinate, furfural and humins. Catalysis Today, 2018, 302 2-15.

[13] Sevilla M, Fuertes A B. Sustainable porous carbons with a superior performance for CO_2 capture. Energy and Environmental Science, 2011, 4 (5): 1765.

[14] Fuertes A B, Sevilla M. Superior capacitive performance of hydrochar-based porous carbons in aqueous electrolytes. ChemSusChem. 2015, 8 (6): 1049-1057.

[15] Wang Q, Li H, Chen L Q, Huang X J. Monodispersed hard carbon spherules with uniform nanopores. Carbon, 2001, 39: 2211-2214.

[16] Wei L, Sevilla M, Fuertes A B, Mokaya R, Yushin G. Hydrothermal carbonization of abundant renewable natural organic chemicals for high-performance supercapacitor electrodes. Advanced Energy Materials, 2011, 1 (3): 356-361.

[17] Kang S M, Ye J, Zhang Y, Chang J. Preparation of biomass hydrochar derived sulfonated catalysts and their catalytic effects for 5-hydroxymethylfurfural production. Rsc Advances, 2013, 3 (20): 7360-7366.

[18] Titirici M M, White R J, Falco C, Sevilla M. Black perspectives for a green future: hydrothermal carbons for environment protection and energy storage. Energy and Environmental Science, 2012, 5 (5): 6796.

[19] Van Zandvoort I, Wang Y H, Rasrendra C B, Van Eck E R H, Bruijnincx P C A, Heeres H J, Weckhuysen B M. Formation, molecular structure, and morphology of humins in biomass conversion: influence of feedstock and processing conditions. ChemSusChem. 2013, 6 (9): 1745-1758.

[20] Van Zandvoort I, Koers E J, Weingarth M, Bruijnincx P C A, Baldus M, Weckhuysen B M. Structural characterization of ^{13}C-enriched humins and alkali-treated ^{13}C humins by 2D solid-state NMR. Green Chemistry, 2015, 17 (8): 4383-4392.

[21] Sumerskii I V, Krutov S M, Zarubin M Y. Humin-like substances formed under the conditions of industrial hydrolysis of wood. Russian Journal of Applied Chemistry, 2010, 83 (2): 320-327.

[22] Herzfeld J, Rand D, Matsuki Y, Daviso E, Mak-Jurkauskas M, Mamajanov I. Molecular structure of humin and melanoidin via solid state NMR. Journal of Physical Chemistry B, 2011, 115 (19): 5741-5745.

[23] Wang S R, Lin H Z, Zhao Y, Chen J P, Zhou J S. Structural characterization and pyrolysis behavior of humin by-products from the acid-catalyzed conversion of C_6 and C_5 carbohydrates. Journal of Analytical and Applied Pyrolysis, 2016, 118: 259-266.

[24] Baccile N, Laurent G, Babonneau F, Fayon F, Titirici M M, Antonietti M. Structural characterization of hydrothermal carbon spheres by advanced solid-state MAS ^{13}C NMR Investigations. The Journal of Physical Chemistry C, 2009, 113 (22): 9644-9654.

[25] Sevilla M, Fuertes A B. The production of carbon materials by hydrothermal carbonization of cellulose. Carbon, 2009, 47 (9): 2281-2289.

[26] Falco C, Perez Caballero F, Babonneau F, Gervais C, Laurent G, Titirici M M, Baccile N. Hydrothermal carbon from biomass: structural differences between hydrothermal and pyrolyzed carbons via 13C solid state NMR. Langmuir, 2011, 27 (23): 14460-14471.

[27] Titirici M M, Antonietti M, Baccile N. Hydrothermal carbon from biomass: a comparison of the local structure from poly-to monosaccharides and pentoses/hexoses. Green Chemistry, 2008, 10 (11): 1204-1212.

[28] Cheng B G, Wang X H, Lin Q X, Zhang X, Meng L, Sun R C, Xin F X, Ren J L. New understandings of the relationship and initial formation mechanism for pseudo-iignin, humins, and acid-induced hydrothermal carbon. Journal of Agricultural and Food Chemistry, 2018, 66 (45): 11981-11989.

[29] Jung D, Zimmermann M, Kruse A. Hydrothermal carbonization of fructose: growth mechanism and kinetic model. Acs Sustainable Chemistry and Engineering. 2018, 6 (11): 13877-13887.

[30] García-Bordejé E, Pires E, Fraile J M. Parametric study of the hydrothermal carbonization of cellulose and effect of acidic conditions. Carbon, 2017, 123: 421-432.

[31] Latham K G, Jambu G, Joseph S D, Donne S W. Nitrogen doping of hydrochars produced hydrothermal treatment of sucrose in H_2O, H_2SO_4, and NaOH. Acs Sustainable Chemistry and Engineering, 2014, 2 (4): 755-764.

[32] Falco C, Baccile N, Titirici M M. Morphological and structural differences between glucose, cellulose and lignocellulosic biomass derived hydrothermal carbons. Green Chemistry, 2011, 13 (11): 3273.

[33] Hu B, Wang K, Wu L H, Yu S H, Antonietti M, Titirici M M. Engineering carbon materials from the hydrothermal carbonization process of biomass. Advanced Materials, 2010, 22 (7): 813-828.

[34] Patil S K R, Lund C R F. Formation and growth of humins via aldol addition and condensation during acid-catalyzed conversion of 5-hydroxymethylfurfural. Energy and Fuels, 2011, 25 (10): 4745-4755.

[35] Patil S K R, Heltzel J, Lund C R F. Comparison of structural features of humins formed catalytically from glucose, fructose and 5-hydroxymethylfurfuraldehyde. Energy and Fuels, 2012, 26 (8): 5281-5293.

[36] Dee S J, Bell A T. A study of the acid-catalyzed hydrolysis of cellulose dissolved in ionic liquids and the factors influencing the dehydration of glucose and the formation of humins. ChemSusChem, 2011, 4 (8): 1166-1173.

[37] Tsilomelekis G, Orella M J, Lin Z, Cheng Z, Zheng W, Nikolakis V, Vlachos D G. Molecular structure, morphology and growth mechanisms and rates of 5-hydroxymethyl furfural (HMF)

derived humins. Green Chemistry, 2016, 18 (7): 1983-1993.

[38] J A, Kumalaputri, Randolph C, Edwin Otten, Heeres H J, Deuss P J. Lewis acid catalyzed conversion of 5-hydroxymethylfurfural to 1, 2, 4-benzenetriol, an overlooked biobased compound. Acs Sustainable Chemistry and Engineering. 2018, 6 (3): 3419-3425.

[39] Hronec M, Fulajtarova K, Liptaj T. Effect of catalyst and solvent on the furan ring rearrangement to cyclopentanone. Applied Catalysis A: General, 2012, 437: 104-111.

[40] Hronec M, Fulajtarova K. Selective transformation of furfural to cyclopentanone. Catalysis Communications, 2012, 24: 100-104.

[41] Horvat J, Klaic B, Metelko B, Sunjic V. Mechanism of levulinic acid formation. Tetrahedron Letters, 1985, 26 (17): 2111-2114.

[42] Akien G R, Qi L, Horvath I T. Molecular mapping of the acid catalysed dehydration of fructose. Chemical Communications, 2012, 48 (47): 5850-5852.

[43] Yong G, Zhang Y G, Ying J Y. Efficient catalytic system for the selective production of 5-hhydroxymethylfurfural from glucose and fructose. Angewandte Chemie-International Edition, 2008, 47 (48): 9345-9348.

[44] Zhao H B, Holladay J E, Brown H, Zhang Z C. Metal chlorides in ionic liquid solvents convert sugars to 5-hydroxymethylfurfural. Science, 2007, 316 (5831): 1597-1600.

[45] Tolborg S, Meier S, Sadaba I, Elliot S G, Kristensen S K, Saravanamurugan S, Riisager A, Fristrup P, Skrydstrup T, Taarning E. Tin-containing silicates: identification of a glycolytic riathway via 3-deoxyglucosone. Green Chemistry, 2016, 18 (11): 3360-3369.

[46] Chen H S, Wang A, Sorek H, Lewis J D, Roman-Leshkov Y, Bell A T. Production of hydroxyl-rich acids from xylose and glucose using Sn-BEA zeolite. Chemistryselect, 2016, 1 (14): 4167-4172.

[47] Elliot S G, Andersen C, Tolborg S, Meier S, Sádaba I, Daugaard A E, Taarning E. Synthesis of a novel polyester building block from pentoses by tin-containing silicates. Rsc Advances, 2017, 7 (2): 985-996.

[48] Sølvhøj A, Taarning E, Madsen R. Methyl vinyl glycolate as a diverse platform molecule. Green Chemistry, 2016, 18 (20): 5448-5455.

[49] Dusselier M, Wouwe P V, Clippel F D, Dijkmans J, Gammon D W, Sels B F. Mechanistic insight into the conversion of tetrose sugars to novel α-hydroxy acid platform molecules, ChemCatChem, 2012, 5 (2): 569-575.

[50] Holm M S, Pagán-Torres Y J, Saravanamurugan S, Riisager A, Dumesic J A, Taarning E. Sn-Beta catalysed conversion of hemicellulosic sugars. Green Chemistry, 2012, 14 702-706

[51] Dusselier M, Van Wouwe P, De Clippel F, Dijkmans J, Gammon D W, Sels B F. Mechanistic insight into the conversion of tetrose sugars to novel α-hydroxy acid platform molecules. ChemCatChem, 2013, 5 (2): 569-575.

[52] Dusselier M, De Clercq R, Cornelis R, Sels B F. Tin triflate-catalyzed conversion of cellulose to valuable (α-hydroxy-) esters. Catalysis Today, 2017, 279 339-344.

[53] Clercq R D, Dusselier M, Christiaens C, Dijkmans J, Iacobescu R I, Pontikes Y, Sels B F. Confine-

[54] Knill C J, Kennedy J F. Degradation of cellulose under alkaline conditions. Carbohydrate Polymers, 2003, 51 (3): 281-300.

[55] Feather M S, Harris J F. Dehydration reactions of carbohydrates. Advances in Carbohydrate Chemistry and Biochemistry Series, 1973, 28: 161-224.

[56] Wang Y L, Deng W P, Wang B J, Zhang Q H, Wan X Y, Tang Z C, Wang Y, Zhu C, Cao Z X, Wang G C, Wan H L. Chemical synthesis of lactic acid from cellulose catalysed by lead (II) ions in water. Nature Communications, 2013, 4: 1-7.

[57] Girisuta B, Janssen L, Heeres H J. Kinetic study on the acid-catalyzed hydrolysis of cellulose to levulinic acid. Industrial and Engineering Chemistry Research. 2007, 46 (6): 1696-1708.

[58] Hu X, Li C Z. Levulinic esters from the acid-catalysed reactions of sugars and alcohols as part of a biorefinery. Green Chemistry, 2011, 13 (7): 1676-1679.

[59] Hu X, Lievens C, Larcher A, Li C Z. Reaction pathways of glucose during esterification: Effects of reaction parameters on the formation of humin type polymers. Bioresource Technology, 2011, 102 (21): 10104-10113.

[60] Zhang J, Li J B, Wu S B, Liu Y. Advances in the catalytic production and utilization of sorbitol. Industrial and Engineering Chemistry Research, 2013, 52 (34): 11799-11815.

[61] Önal Y, Schimpf S, Claus P. Structure sensitivity and kinetics of D-glucose oxidation to D-gluconic acid over carbon-supported gold catalysts. Journal of Catalysis, 2004, 223 (1): 122-133.

[62] Tan X S, Deng W P, Liu M, Zhang Q H, Wang Y. Carbon nanotube-supported gold nanoparticles as efficient catalysts for selective oxidation of cellobiose into gluconic acid in aqueous medium. Chemical Communications, 2009, 41 (46): 7179-7181.

[63] Yamaguchi A, Hiyoshi N, Sato O, Shirai M. Sorbitol dehydration in high temperature liquid water. Green Chemistry, 2011, 13 (4): 873-881.

[64] Goossen L J, Rudolphi F, Oppel C, Rodriguez N. Synthesis of ketones from α-oxocarboxylates and aryl bromides by Cu/Pd-catalyzed decarboxylative cross-coupling. Angewandte Chemie—International Edition, 2008, 47 (16): 3043-3045.

[65] Wang H, Guo L N, Duan X H. Decarboxylative acylation of cyclic enamides with α-oxocarboxylic acids by palladium-catalyzed C—H activation at room temperature. Organic Letters, 2012, 14 (17): 4358-4361.

[66] Dusselier M, Sels B F. Selective catalysis for cellulose conversion to lactic acid and other α-hydroxy acids. Selective Catalysis for Renewable Feedstocks and Chemicals. 2014, 353: 85-125.

[67] Jadhav H, Pedersen C M, Solling T, Bols M. 3-deoxy-glucosone is an intermediate in the formation of furfurals from D-glucose. ChemSusChem, 2011, 4 (8): 1049-1051.

[68] Rasrendra C B, Windt M, Wang Y, Adisasmito S, Makertihartha I G B N, Van Eck E R H, Meier D, Heeres H J. Experimental studies on the pyrolysis of humins from the acid-catalysed dehydration of C_6-sugars. Journal of Analytical and Applied Pyrolysis, 2013, 104: 299-307.

[69] Ji N, Zhang T, Zheng M Y, Wang A Q, Wang H, Wang X D, Chen J G. Direct catalytic con-

version of cellulose into ethylene glycol using nickel-promoted tungsten carbide catalysts. Angewandte Chemie—International Edition, 2008, 47 (44): 8510-8513.

[70] Wang A Q, Zhang T. One-pot conversion of cellulose to ethylene glycol with multifunctional tungsten-based catalysts. Accounts of Chemical Research, 2013, 46 (7): 1377-1386.

[71] Liu Y, Luo C, Liu H C. Tungsten Trioxide Promoted Selective Conversion of cellulose into propylene glycol and ethylene glycol on a ruthenium catalyst. Angewandte Chemie—International Edition, 2012, 51 (13): 3249-3253.

[72] Onda A, Ochi T, Kajiyoshi K, Yanagisawa K. A new chemical process for catalytic conversion Of D-glucose into lactic acid and gluconic acid. Applied Catalysis A: General, 2008, 343 (1-2): 49-54.

第五章

葡萄糖及生物质衍生呋喃类物质水热转化生成的可溶性副产物鉴定

5.1 引言

根据目标产物不同，木质纤维素水热炼制技术可以分为水热碳化技术和水热液化技术。木质纤维素水热液化技术以获取液体燃料或高附加值有机化学品为目标，在这一过程中希望尽可能地抑制水热焦炭的生成；木质纤维素水热碳化技术以获取水热焦炭为目标，在这一过程中希望尽可能地促进水热焦炭的生成。因此，阐明水热焦炭的生成机理和碳水化合物的水热转化路径，对于木质纤维素水热液化技术和水热碳化技术都具有重要意义。

在第四章的研究中发现，只有在水热条件下生成α-羰基醛和α-羰基酸的含氧有机物才能在水热转化过程中形成水热焦炭，而碳水化合物在水热转化过程中易生成α-羰基醛，因此笔者提出碳水化合物在水热转化过程中形成的α-羰基醛，如丙酮醛、3-脱氧葡萄糖醛酮、3-脱氧木糖醛酮和 2,5-二氧代-6-羟基己醛等，是形成水热焦炭的主要前驱体，而羟醛缩合反应是这些α-羰基醛发生聚合的关键步骤（图 5-1）[1-3]。

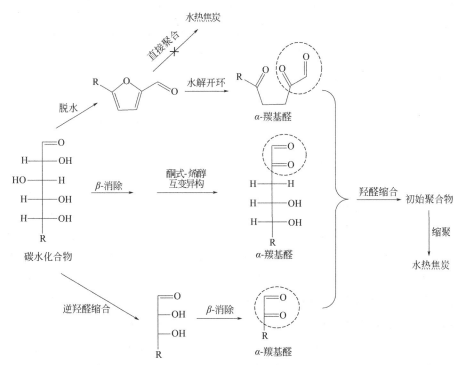

图 5-1 碳水化合物生成α-羰基醛和水热焦炭的路径（R 表示 H 或 CH$_2$OH）

第四章的研究发现，葡萄糖、HMF、糠醛和糠醇在水热碳化过程中会形成 40%～60%的水溶性有机物[3-9]。这些水溶性副产物可能是上述模型化合物生成水热焦炭的关键中间体，或者是在水热焦炭形成过程中释放出来的副产品[3,7]。因此，对这些水溶性化合物的分子结构进行鉴定，将有助于揭示葡萄糖的水热转化路径和水热焦炭的形成机理。然而，到目前为止，对这些水溶性化合物（特别是分子量大于 126 的化合物）的认识还很匮乏。这是因为含有大量羟基、羰基和羧基官能团的有机物通常具有挥发性差、极性高、热稳定性差等特征，导致很难用 GC-MS 对这些物质的结构进行分析，而 GC-MS 是鉴定复杂混合物中挥发性有机物的常用方法。为了能够采用 GC-MS 对碳水化合水热转化生成的水溶性副产物进行鉴定，Poerschmann 等通过硅烷基化、甲基化和酰化等各种衍生化手段来提高这些化合物的挥发性，从而利用 GC-MS 鉴定出了多种分子量在 120～300 范围的呋喃类化合物和羟基苯并呋喃衍生物[10]。但是，这种分析方法过程烦琐且结果可靠性不高。

另一方面，LC-MS^n 是鉴定低挥发性、高极性水溶性化合物的有效方法，已被大量的研究人员用于分析生物质分解过程中形成的水溶性化合物[7,11-16]。Maruani 等利用 LC-MS 和 LC-MS^2 鉴定了葡萄糖在酸性介质中水热脱水过程中形成的一些水溶性低聚糖，并提出这些低聚糖可能参与水热焦炭的形成[7]。Rasmussen 等采用 LC-MS 和 LC-MS^2 鉴定出木质纤维素类生物质在水热处理过程中可生成酚类化合物、4-羟基-3-甲氧基肉桂酸酰化的双环戊糖衍生物等高极性和低挥发性的含氧有机物[13,14]。然而，利用 LC-MS 和 LC-MS^2 对葡萄糖及其衍生物在水热转化过程中形成的水溶性副产物的鉴定还远远不够，因为甚至没有文献报道利用 LC-MS 和 LC-MS^2 检测到已经被 GC-MS 通过衍生化反应鉴定的呋喃类化合物和羟基苯并呋喃衍生物[10]。

在本章中，笔者用 LC-MS 和 LC-MS^2 分析了葡萄糖、HMF、糠醛和糠醇在水热碳化过程中形成的水溶性化合物。在葡萄糖的水热转化产物中，除已报道的 3-脱氧葡萄糖醛酮、HMF、乳酸和乙酰丙酸等化合物外，还鉴定出多种分子量在 170～270 之间、含有羟基和羰基的呋喃类化合物和碳环含氧有机物。在 HMF、糠醛和糠醇的水热转化产物中也同样鉴定出大量分子量在 150～272 之间的、含有羟基和羰基的碳环类物质。对所检测到的碳环类含氧有机物的结构分析表明，这些物质的生成过程中涉及碳链增长反应和碳链断裂反应，因此笔者提出，这些碳环含氧有机物是由这些模型化合物在水热转化过程中生成的 α-羰基醛经羟醛缩合生成含有碳环结构的初级聚合物，进而发生 C—C 键断裂而形成。通过对模型化合物在水热条件下所生成的 α-羰基醛的聚合路径进行分析，笔者发现 α-羰基醛都可以通过羟醛缩合、缩醛环化、水解开环、分子内羟醛缩合、C—C 键断裂等一系列反应生成六元碳环化合物。

5.2 LC-MS^n 对可溶性葡萄糖水热转化产物的分析条件

用于 LC-MS 和 LC-MS^2 分析的样品制备：将 30mL 水和含有 0.12mol 碳原子的反应原料（葡萄糖、HMF、糠醛和糠醇）置于内衬聚四氟乙烯的高压水热合成釜中，在 220℃下保持一定时间进行水热碳化。水热碳化过程结束后，过滤，将收集的滤液在 50℃下蒸发除去大部分的水溶剂得到黏稠的褐色样品，然后将样品用甲醇稀释 20 倍，经 0.22μm 微孔滤膜过滤后进行 LC-MS 分析，以纯甲醇作为空白对照。

LC-MS 和 LC-MS^2 分析在配备了质谱仪（Thermo Science Q Exactive，美国）的超高性能液相色谱（UHPLC-Q）仪器（Dinonex Ultimate 3000，美国）上进行，电离方式为电加热喷雾电离（HESI）。使用制造商提供的混合物在 m/z 为 50～3200 范围内校准质量轴。正、负模式下 MS^2 的碰撞诱导解离能分别为 4.0kV 和 3.2kV。

色谱柱为 Eclipse Plus C$_{18}$ 250mm×4.6mm×5μm。柱温 25℃。流动相分别为含 0.1%甲酸的水溶液和纯甲醇。色谱柱的梯度洗脱条件见表 5-1。

表 5-1　色谱柱的梯度洗脱条件

时间/min	流速/(mL/min)	0.1%甲酸/%	纯甲醇/%
00.00	0.60	95	5
03.00	0.60	95	5
20.00	0.60	50	50
25.00	0.60	5	95
30.00	0.60	5	95
33.00	0.60	95	5
40.00	0.30	95	5

5.3 LC/MS 和 LC/MS^2 谱图的分析过程

以纯甲醇测试谱图作为背景，通过工作站软件对样品谱图进行背景扣除。然后再进行下一步的成分筛选。由于 LC-MS 测试成分复杂，目标成分不明确，所以仅对主要离子的一级、二级质谱进行统计；另外，由于样品的总离子流图（TIC）中，背景干扰较大（干扰离子主要来自溶剂和流动相），所以选择利用基峰（basepeak，质谱图中的最强离子）谱图进行成分筛选。

根据 MS 质谱数据确定了化合物的精确分子量和化学式。在正电离模式下，化合物可产生 $[M+H]^+$、$[M+H_2O]^+$、$[M+Na]^+$ 和 $[M+K]^+$ 等准分子离子峰，有助于测定化合物的精确质量。例如，葡萄糖降解产物中保留时间为 16.50min 的化合物在 m/z 为 173.08、190.11 和 195.06 处有三个离子峰（图 5-2），分别对应于 $[M+H]^+$、$[M+H_2O]^+$ 和 $[M+Na]^+$，因此确定该化合物的精确分子量为 172.07。

在负电离模式的谱图中，化合物可产生 $[M-H]^-$、$[M+Cl]^-$ 和 $[M-H+HCOOH]^-$ 等准分子离子峰。在这些准分子离子峰中，特征同位素团簇的存在，如典型的 $[M+Cl]^-$ 能够辅助确定待分析的物质的分子量[17]。例如，葡萄糖降解产物中保留时间为 4.89min 的化合物的质谱图中出现 197.02 和 199.02 两个离子峰（图 5-3），199.02 的峰强度约为 197.02 的 1/3，表明 197.02 处的峰是 $[M+Cl]^-$ 的准分子离子峰。此外，在 161.04 处还观察到一个明显的峰，该峰是 $[M-H]^-$ 的准分子离子峰。因此，可以确定该化合物的分子量为 162.05。

众所周知，葡萄糖、HMF、糠醛、糠醇和水都只含有 C、H 和 O 三种元素，而 H、C 和 O 的精确原子量分别为 1.0078、12 和 15.9949。因此，根据质谱中离子的 m/z，可以计算出离子所对应的化学式。根据上述原理，笔者开发了一款能够专用于解析含有 C、H、O 三种元素有机物的高分辨质谱解析软件。该软件根据下面几个不等式对有机物的质谱进行解析，确定化合物的质谱图中每一个质谱峰所对应的离子的化学式。

$$m_2 = 12C + 1.0078H + 15.9949O$$

$$|m_2 - m| < 0.01$$

$$C - 4 \leqslant H \leqslant 2C + 3$$

$$C \geqslant O - 2$$

上述不等式组中，m 为实验所检测到的离子的质量，m_2 为候选离子的理论质量，C、H、O 分别为离子所含有的碳、氢、氧原子的数目。质谱离子峰解析算法如图 5-4 所示。该软件通过对 m 取整得到整数 m_1，然后通过 $H = m_1 - 12C - 16O$ 计算出候选离子的化学式，进而计算出该候选离子的理论质量，然后结合高分辨质谱的分辨率确定候选离子的理论质量与检测到的离子质量的差值阈值为 0.01，并比较候选离子的质量与检测到的离子的质量之差是否在阈值范围内。在计算初始确定 C、O 的值均为 0，通过每次循环 C 数量增加和 O 数量增加，不停地对所有可能的离子进行筛查，最终确定出符合条件的离子的化学式。

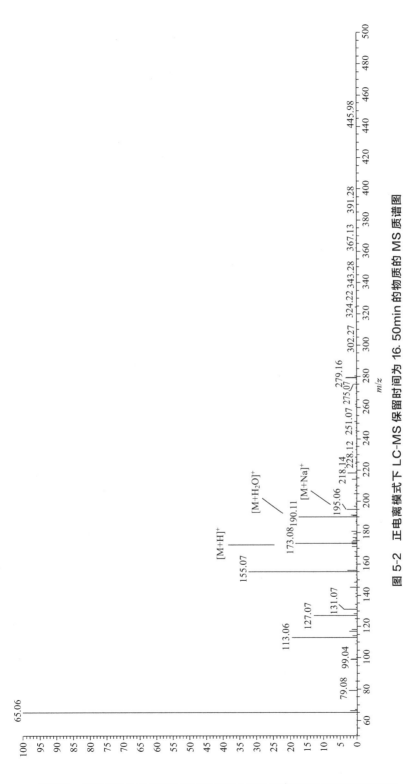

图 5-2 正电离模式下 LC-MS 保留时间为 16.50min 的物质的 MS 质谱图

第五章 葡萄糖及生物质衍生呋喃类物质水热转化生成的可溶性副产物鉴定

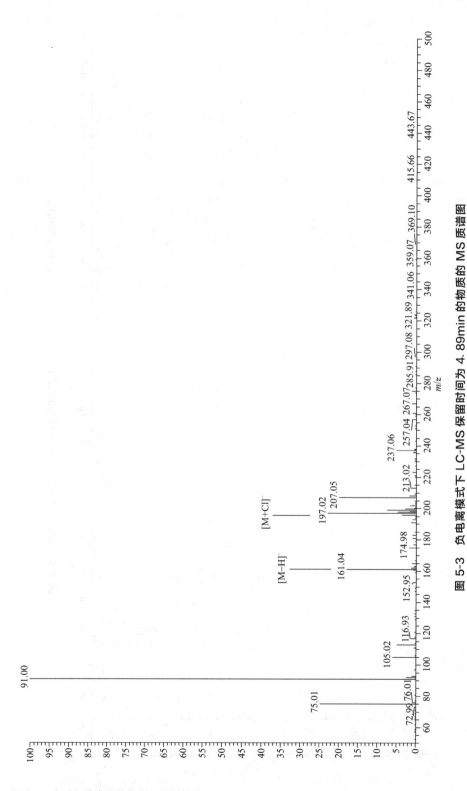

图 5-3 负电离模式下 LC-MS 保留时间为 4.89min 的物质的 MS 质谱图

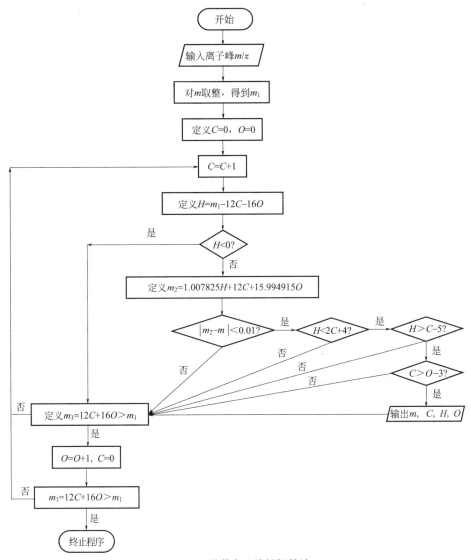

图 5-4 质谱离子峰解析算法

5.4 葡萄糖水热转化生成的水溶性物质鉴定

5.4.1 LC-MS 确定葡萄糖降解生成的水溶性化合物的化学式

图 5-5 展示了葡萄糖降解生成的水溶性化合物样品在正电离模式和负电离模式下的 LC-MS 基峰谱图。很明显,在正电离和负电离模式下的 LC-MS 基

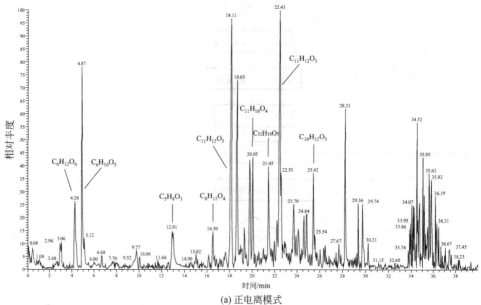

(a) 正电离模式

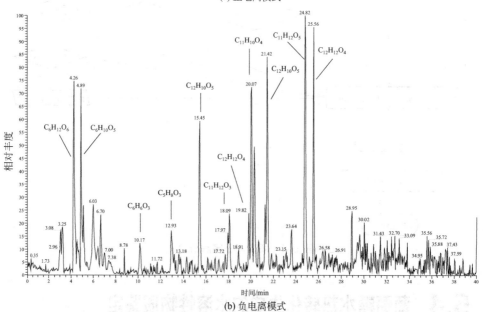

(b) 负电离模式

图 5-5 葡萄糖降解产物的 LC-MS 基峰谱图

峰谱图均出现多个峰，表明葡萄糖在降解过程中生成了大量的水溶性化合物。表 5-2 列出了所有已鉴定化合物的精确分子量和相应的分子式。显然，在正、负电离模式下，LC-MS 均检测到分子式为 $C_6H_{12}O_6$（4.26min）、$C_6H_{10}O_5$（4.89min）、$C_5H_8O_3$（12.93min）、$C_{11}H_{12}O_3$（18.11min）、$C_{12}H_{12}O_4$

（19.82min）、$C_{11}H_{10}O_4$（20.07min）、$C_{12}H_{10}O_5$（21.42min）和 $C_{11}H_{12}O_5$（24.82min）的化合物，而其他化合物则只能在正离子模式或负离子模式下被检测到。因此，不能根据谱图的峰面积来对这些化合物进行定量分析。在所有检测到的化合物中，除分子式为 $C_6H_{12}O_6$、$C_6H_{10}O_5$、$C_5H_8O_3$ 和 $C_8H_{12}O_4$ 的化合物外，其余化合物的双键当量（DBE）均不低于4，表明这些化合物中存在大量的不饱和基团，如呋喃环结构或苯环结构[14]。

表 5-2　LC-MS 检测到的葡萄糖降解生成的可溶性物质的分子式及其特征

保留时间/min	分子式	DBE	$(H-2O)/C$	理论分子量	检测到的$[M+H]^+$	检测到的$[M-H]^-$	电离模式
4.28	$C_6H_{12}O_6$	1	0	180.05	181.06	179.05	正/负
4.89	$C_6H_{10}O_5$	2	0	162.05	163.04	161.04	正/负
10.17	$C_6H_6O_3$	4	0	126.03	—	125.02	负
12.91	$C_5H_8O_3$	2	0.4	116.04	117.05	115.04	正/负
15.45	$C_{12}H_{10}O_5$	8	0	234.05	—	233.04	负
16.50	$C_8H_{12}O_4$	3	0.5	172.07	173.08	—	正
17.97	$C_{11}H_{16}O_7$	4	0.18	260.08	—	259.08	负
18.11	$C_{11}H_{12}O_3$	6	0.55	192.07	193.08	191.07	正/负
19.82	$C_{12}H_{12}O_3$	7	0.33	220.07	221.08	219.06	正/负
20.07	$C_{11}H_{10}O_4$	7	0.18	206.05	207.06	205.05	正/负
21.42	$C_{12}H_{10}O_5$	8	0	234.05	235.07	233.04	正/负
22.43	$C_{11}H_{12}O_3$	6	0.55	192.07	193.09	—	正
23.64	$C_{12}H_{10}O_4$	8	0.17	218.05	—	217.05	负
24.82	$C_{11}H_{12}O_5$	10	0.18	224.07	225.07	223.06	正/负
25.42	$C_{10}H_{12}O_3$	5	0.6	180.07	181.09	—	正
25.56	$C_{12}H_{12}O_3$	7	0.33	220.07	—	219.07	负

$(H-2O)/C$ 的值表示化合物中每个碳原子所分摊的净氢原子数。这个值只有在发生氧化还原反应时才会发生变化。众所周知，碳水化合物在水热碳化过程中会发生脱水、水合、缩醛、醚化、酯化、羟醛缩合、逆羟醛缩合和 C—C 键水解断裂等多种反应，其中，仅涉及水分子的加成或脱除的反应，如脱水、水合、缩醛、醚化、酯化、羟醛缩合和逆羟醛缩合等都不属于氧化还原反应，所以这些反应都不会导致产物的 $(H-2O)/C$ 值相对于反应物的 $(H-2O)/C$ 值发生变化。相反，C—C 键水解断裂反应可以认为是一种分子内的氧化还原反应，它会生成一个 $(H-2O)/C$ 值低于反应物的氧化产物和一个 $(H-2O)/C$ 值高于反应物的还原产物。因为葡萄糖和水的 $(H-2O)/C$ 值均为0，故除非发生 C—C 键断裂反应，否则葡萄糖在水热转化过程中生成的化

合物的 $(H-2O)/C$ 应保持为 0。例如，葡萄糖在水热条件下生成乳酸、HMF 和 1,2,4-苯三酚这些产物时都只涉及水分子的加成和脱除反应[4,6,18,19]，而这些化合物的 $(H-2O)/C$ 值均为 0。反之，葡萄糖生成乙酰丙酸和甲酸的反应涉及 C—C 键水解断裂反应[29]，而乙酰丙酸和甲酸的 $(H-2O)/C$ 值都与葡萄糖不同。因此，笔者认为，根据所鉴定的化合物的 $(H-2O)/C$ 值可以判断该化合物的形成过程中是否涉及 C—C 键水解断裂反应。表 5-2 显示了这些已鉴定化合物的 $(H-2O)/C$ 值。对于化学式为 $C_6H_{12}O_6$、$C_6H_{10}O_5$、$C_6H_6O_3$ 和 $C_{12}H_{10}O_5$ 的化合物，它们的 $(H-2O)/C$ 的值均为 0，表明从葡萄糖中生成这些化合物的反应不涉及 C—C 断裂反应。相反，其他化合物的 $(H-2O)/C$ 值均不为 0，表明这些化合物的形成过程中必然发生了 C—C 键断裂反应。

5.4.2 根据 MS^2 质谱图确定化合物的分子结构

MS^2 质谱图可以提供官能团的信息，从而辅助确定化合物的核心结构和官能团。特别地，MS^2 谱中 m/z 在 70~130 范围内的碎片离子通常只对应一个或几个特殊的化学结构。例如，正离子模式下检测到的 m/z 为 95.05 的碎片离子的化学式为 $[C_6H_7O]$，与此化学式相对应的是质子化的苯酚。这些特征碎片离子可以辅助确定相应化合物的核心结构。附图 1-15 显示了所有这些检测到的化合物的 MS^2 质谱图及其可能的解离路径。应该注意的是，每个化学式都有很多异构体，所以无法根据 MS^2 质谱图的信息确定分子的确切结构。

如表 5-3 所示，m/z 为 71.01 和 89.02 的碎片离子的化学式分别为 $[C_3H_3O_2]$ 和 $[C_3H_5O_3]$，而这些化学式所对应的结构都属于链式醛/酸。这些碎片离子在 $C_6H_{12}O_6$(4.28min)、$C_6H_{10}O_5$(4.89min)、$C_{11}H_{16}O_7$(17.97min) 和 $C_{12}H_{12}O_4$(25.56min) 的 MS^2 质谱图中的存在，表明这些化合物含有链式醛/酸结构。因此，化合物 $C_6H_{12}O_6$(4.28min) 应为葡萄糖，而化合物 $C_6H_{10}O_5$(4.89min) 应当为葡萄糖经 β-消除反应和酮式-烯醇互变异构反应生成的 3-脱氧葡萄糖醛酮[2,21]。

表 5-3 几种典型碎片离子峰的化学式及其结构

碎片离子的 m/z	碎片离子的化学式	电离模式	碎片离子种类	含有该碎片离子的物质
71.01	$[C_3H_3O_2]$	负	链式醛/酸	$C_6H_{12}O_6$(4.28min),$C_6H_{10}O_5$(4.89min)
89.02	$[C_3H_5O_3]$	负	链式醛/酸	$C_6H_{12}O_6$(4.28min),$C_6H_{10}O_5$(4.89min),$C_{11}H_{16}O_7$(17.97min)

续表

碎片离子的 m/z	碎片离子的化学式	电离模式	碎片离子种类	含有该碎片离子的物质
91.05	$[C_7H_7]$	正	六元碳环	$C_{11}H_{12}O_3$(18.11min),$C_{12}H_{12}O_4$(19.81min),$C_{11}H_{10}O_4$(20.07min),$C_{11}H_{12}O_3$(22.43min)
93.03	$[C_6H_5O]$	负	六元碳环	$C_{12}H_{10}O_4$(23.64min)
93.07	$[C_7H_9]$	正	六元碳环	$C_{11}H_{12}O_3$(18.11min),$C_{11}H_{12}O_3$(22.43min),
95.05	$[C_6H_7O]$	正	六元碳环	$C_8H_{12}O_4$(16.50min),$C_{11}H_{12}O_3$(18.11min),$C_{12}H_{12}O_4$(19.81min)
105.07	$[C_8H_9]$	正	六元碳环	$C_{11}H_{12}O_3$(18.11min),$C_{12}H_{12}O_4$(19.81min),$C_{11}H_{10}O_4$(20.07min),$C_{11}H_{12}O_3$(22.43min)
109.06	$[C_7H_9O]$	正	六元碳环	$C_8H_{12}O_4$(16.50min),$C_{11}H_{12}O_3$(18.11min),$C_{11}H_{10}O_4$(20.07min),$C_{11}H_{12}O_3$(22.43min)
119.09	$[C_9H_{11}]$	正	六元碳环	$C_{11}H_{12}O_3$(18.11min),$C_{11}H_{12}O_3$(22.43min)
131.09	$[C_{10}H_{11}]$	正	六元碳环	$C_{11}H_{12}O_3$(22.43min)
133.06	$[C_9H_9O]$	正	六元碳环	$C_{12}H_{12}O_4$(19.81min),$C_{11}H_{10}O_4$(20.07min)
147.08	$[C_{10}H_{11}O]$	正/负	六元碳环	$C_{11}H_{12}O_3$(18.11min),$C_{12}H_{12}O_4$(19.81min),$C_{11}H_{12}O_3$(22.43min),$C_{12}H_{12}O_4$(25.56min)
99.04	$[C_5H_7O_2]$	正/负	呋喃环	$C_{11}H_{16}O_7$(17.97min),$C_{10}H_{12}O_3$(25.42min)
107.01	$[C_6H_3O_2]$	负	呋喃环	$C_6H_6O_3$(10.17min),$C_{11}H_{12}O_5$(24.82min),
109.03	$[C_6H_5O_2]$	正/负	呋喃环	$C_{12}H_{10}O_5$(15.45min),$C_{12}H_{10}O_5$(21.45min),
123.01	$[C_6H_3O_3]$	负	呋喃环	$C_6H_6O_3$(10.17min),$C_{12}H_{10}O_5$(15.45min),$C_{12}H_{12}O_4$(19.81min),$C_{12}H_{10}O_4$(23.64min)
125.02	$[C_6H_5O_3]$	负	呋喃环	$C_6H_6O_3$(10.17min),$C_{12}H_{10}O_5$(15.45min),$C_{11}H_{16}O_7$(17.97min),$C_{12}H_{12}O_4$(19.81min),$C_{11}H_{12}O_4$(24.82min)

m/z 为 91.05、93.03、93.07、95.05、105.07、109.06、119.09、131.09、133.06 和 147.08 的碎片离子对应的化学式分别为 $[C_7H_7]$、$[C_6H_5O]$、$[C_7H_9]$、$[C_6H_7O]$、$[C_8H_9]$、$[C_7H_9O]$、$[C_9H_{11}]$、$[C_{10}H_{11}]$、$[C_9H_9O]$ 和 $[C_{10}H_{11}O]$。这些碎片离子的 DBE 值都不低于 3，而它们所含有的氧原子个数都不超过 1，表明所有的这些碎片离子都含有六元碳环结构。这些碎片离子在 $C_8H_{12}O_4$（16.50min）、$C_{11}H_{12}O_3$（18.11min）、$C_{12}H_{12}O_4$（19.81min）、$C_{11}H_{10}O_4$（20.07min）、$C_{11}H_{12}O_3$（22.43min）、$C_{12}H_{10}O_4$（23.64min）和

$C_{12}H_{12}O_4$（25.56min）的 MS^2 质谱图中存在，表明这些化合物含有六元碳环结构。特别地，化学式为 $[C_{10}H_{11}]$、$[C_9H_9O]$ 和 $[C_{10}H_{11}O]$ 的碎片离子的 DBE 值都超过 5，表明这几种碎片离子很可能含有双碳环结构。这几种碎片离子存在于 $C_{11}H_{12}O_3$（18.11min）、$C_{12}H_{12}O_4$（19.81min）、$C_{11}H_{10}O_4$（20.07min）、$C_{11}H_{12}O_3$（22.43min）和 $C_{12}H_{12}O_4$（25.56min）的 MS^2 质谱图中，表明这些物质含有双碳环结构。需要注意的是，在 MS^2 质谱图中检测到电离苯衍生物和电离苯酚的离子峰，并不能证明这些化合物一定含有苯酚结构，因为环己二烯醇衍生物的分子在解离过程中也会产生含有苯环结构的离子碎片。特别地，在 $C_8H_{12}O_4$（16.50min）的 MS^2 质谱图中检测到 95.05 处的碎片离子（对应质子化苯酚），但该化合物的 DBE 值仅为 3，表明该化合物不可能含有酚环结构，因此，这个化合物应含有环己二烯醇结构。此外，$C_{11}H_{12}O_3$（18.11min）和 $C_{12}H_{12}O_4$（19.81min）的 MS^2 质谱图中同时存在质子化甲苯和质子化苯酚，证明这些化合物同样应当含有环己二烯结构而不是苯酚结构。总之，MS^2 质谱图表明，$C_8H_{12}O_4$（16.50min）、$C_{11}H_{12}O_3$（18.11min）、$C_{12}H_{12}O_4$（19.81min）、$C_{11}H_{10}O_4$（20.07min）、$C_{11}H_{12}O_3$（22.43min）、$C_{12}H_{10}O_4$（23.64min）和 $C_{12}H_{12}O_4$（25.56min）均为碳环类物质。

相似地，m/z 为 99.04、107.01、109.03、123.01 和 125.02 的碎片离子对应的化学式分别为 $[C_5H_7O_2]$、$[C_6H_3O_2]$、$[C_6H_5O_2]$、$[C_6H_3O_3]$ 和 $[C_6H_5O_3]$，这些碎片离子的 DBE 值均不低于 3，而它们含有的氧原子数不低于 2，表明这些离子应含有呋喃环结构。这些碎片离子在 $C_6H_6O_3$（10.17min）、$C_{12}H_{10}O_5$（15.45min）、$C_{11}H_{16}O_7$（17.97min）、$C_{12}H_{12}O_4$（19.81min）、$C_{12}H_{10}O_5$（21.45min）、$C_{12}H_{10}O_4$（23.64min）、$C_{11}H_{12}O_5$（24.82min）和 $C_{10}H_{12}O_3$（25.42min）的 MS^2 质谱图中的存在，表明这些化合物都是呋喃类物质。显然，含有呋喃结构的化合物 $C_6H_6O_3$（10.17min）应该是葡萄糖脱水所生成的 HMF[6,22]，而 $C_{12}H_{10}O_5$（15.45min）和 $C_{12}H_{10}O_5$（21.45min）则可能是 HMF 通过缩醛/醚化等反应所生成的二聚体。

另一方面，MS^2 质谱图中碎片离子之间的差异可以显示丢失碎片的质量，从而辅助确定化合物的官能团[16,23]。检测到的质量丢失和相应的官能团列于表 5-4 中。丢失的质量 14.02、18.01、26.02、27.99、28.03、30.01 和 43.99 分别对应着丢失官能团 CH_2、H_2O、C_2H_2、CO、C_2H_4、CH_2O 和 CO_2。这些质量的丢失表明相应官能团存在于被检测的化合物中。

表 5-4 离子碎片间丢失基团的质量及相应的基团

检测到的质量丢失	丢失的基团	丢失基团的理论质量	相应的结构
14.02	CH_2	14.0157	亚甲基
18.01	H_2O	18.0106	羟基
26.02	C_2H_2	26.0157	乙炔基
27.99	CO	27.9949	羰基
28.03	C_2H_4	28.0313	乙烯基
30.01	CH_2O	30.0106	醛基
43.99	CO_2	43.9898	羧基
82.04	C_5H_6O	82.0417	甲基呋喃
94.04	C_6H_6O	94.0417	苯酚
96.02	$C_5H_4O_2$	96.0210	糠醛
98.03	$C_5H_6O_2$	98.0326	糠醇
110.04	$C_6H_6O_2$	110.0366	甲基糠醛

在化合物 $C_6H_{10}O_5$(4.89min)、$C_{11}H_{12}O_3$(18.11min)、$C_{12}H_{12}O_4$(19.81min)、$C_{11}H_{10}O_4$(20.07min)、$C_{11}H_{12}O_3$(22.43min) 和 $C_{12}H_{10}O_4$(23.64min) 的 MS^2 质谱图中,检测到 26.02 的质量丢失(对应于 C_2H_2 基团的丢失)和 28.03 的质量丢失(对应于 C_2H_4 基团的丢失),表明在这些化合物中应该存在脂肪族 C—C 基团。脂肪族 C—C 基团的存在表明化合物 $C_6H_{10}O_5$(4.89min)应当为葡萄糖脱水所生成的 3-脱氧葡萄糖醛酮[2,21]。

在化合物 $C_5H_8O_3$(12.91min)、$C_{11}H_{16}O_7$(17.97min)、$C_{12}H_{10}O_4$(23.64min) 和 $C_{12}H_{12}O_4$(25.56min) 的 MS^2 质谱图中发现 43.99 的质量丢失,这表明这些化合物中存在羧基。羧基的存在进一步证实了化合物 $C_5H_8O_3$(12.91min) 是由 HMF 经水合反应生成的乙酰丙酸[20,24,25]。

质量 82.04、96.02、98.03 和 110.04 的碎片丢失分别归因于解离过程中丢失甲基呋喃、糠醛、糠醇和甲基糠醛。化合物 $C_{12}H_{10}O_5$(15.45min)、$C_{12}H_{10}O_5$(21.45min)、$C_{11}H_{12}O_5$(24.82min) 和 $C_{10}H_{12}O_3$(25.42min) 的 MS^2 质谱中存在这些质量丢失,表明这些化合物的分子离子在解离过程中释放了呋喃基团。对于化合物 $C_{12}H_{10}O_5$(15.45min)、$C_{12}H_{10}O_5$(21.45min)、$C_{11}H_{12}O_5$(24.82min) 和 $C_{10}H_{12}O_3$(25.42min),分子离子失去呋喃基团可进一步生成含有呋喃结构的碎片离子,表明这些化合物均含有两个呋喃环。因此,$C_{12}H_{10}O_5$(15.45min) 或 $C_{12}H_{10}O_5$(21.45min) 应该是 HMF 经醚化反应生成的二聚体(5,5'-双呋喃-2-甲醛)[1,5]。

图 5-6 总结了这些检测到的化合物可能的分子结构。化合物 $C_6H_{12}O_6$（4.28min）、$C_6H_{10}O_5$（4.89min）、$C_6H_6O_3$（10.17min）和 $C_5H_8O_3$（12.91min）分别为葡萄糖、3-脱氧葡萄糖醛酮、羟甲基糠醛和乙酰丙酸。化合物 $C_{12}H_{10}O_5$（15.45min）、$C_{12}H_{10}O_5$（21.42min）、$C_{11}H_{12}O_5$（24.82min）和 $C_{10}H_{12}O_3$（25.42min）均含有呋喃环状结构，而化合物 $C_8H_{12}O_4$（16.50min）、$C_{11}H_{12}O_3$（18.11min）、$C_{12}H_{12}O_4$（19.82min）、$C_{11}H_{10}O_4$（20.07min）、$C_{11}H_{12}O_3$（22.43min）、$C_{12}H_{10}O_4$（23.64min）、$C_{12}H_{10}O_4$（25.56min）是碳环类物质。由于 $C_{12}H_{12}O_4$（19.81min）和 $C_{12}H_{10}O_4$（23.64min）的解离过程中既可以产生含有碳环结构的碎片离子，又可以产生含有呋喃环结构的碎片离子，因此这些化合物既含有碳环结构又含有呋喃环结构。

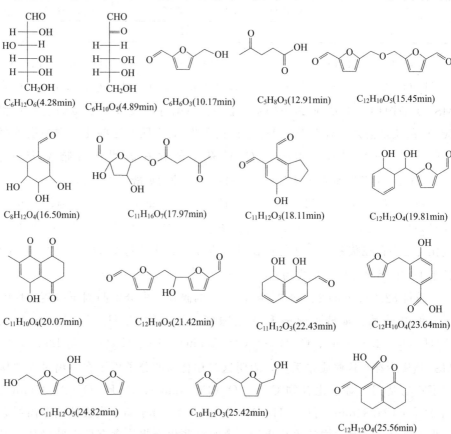

图 5-6 通过 LC-MS² 谱图推测的物质可能的分子结构

给出了各取代基的种类，但没有确定异构体

上图展示的化合物 $C_{11}H_{12}O_3$、$C_{11}H_{10}O_4$、$C_{11}H_{12}O_5$、$C_{12}H_{10}O_4$、$C_{12}H_{12}O_4$ 和 $C_{12}H_{10}O_5$ 的分子结构与 Poerschmann 等提出的羟基苯并呋喃衍生物的分子结构有很大的不同[10]。由于羟基苯并呋喃衍生物分子结构的断裂很难生成离子化的苯类化合物和苯酚，而环己二烯醇衍生物则相反，所以本文提出的分子结构更接近实际。总之，这项工作和 Poerschmann 等报告的工作都证实了葡萄糖的水热转化可以生成多个分子量在 170~220 之间的碳环化合物。

5.4.3 葡萄糖水热转化过程中可溶性碳环类物质的形成机理

上述结果证实了葡萄糖在水热转化过程中可以生成多种 $C_8 \sim C_{12}$ 含氧碳环有机物。这些含氧碳环有机化合物可能是水热焦炭形成的前体，也可能是水热焦炭形成过程中的副产物。因此，分析这些碳环氧基有机化合物的形成机理可以辅助揭示水热焦炭的形成机理。

葡萄糖的碳链长度为 6，但所鉴定的含氧碳环有机物的碳链长度均不少于 8，说明在这些化合物的生成过程中一定涉及了 C—C 偶联反应。众所周知，羟醛缩合反应是醛/酮类化合物碳链增长的常见反应之一，而葡萄糖的降解会产生多种羰基化合物，因此，笔者推测羟醛缩合反应是葡萄糖降解产物实现碳链增长生成这些碳环含氧有机物的关键步骤。另一方面，所检测到的碳环类物质 $(H-2O)/C$ 值与葡萄糖不相同，表明这些化合物的形成涉及 C—C 键断裂反应。因此，这些可溶性碳环类物质由葡萄糖水热转化过程中生成的中间小分子通过 C—C 偶联而形成的初始聚合物再发生 C—C 键断裂所释放出来的。

第四章的研究表明，葡萄糖在水热转化过程中生成的 α-羰基醛，如丙酮醛、3-脱氧葡萄糖醛酮和 2,5-二氧代-6-羟基己醛等，可以发生羟醛缩合反应生成初始聚合物，这些聚合物被认为是水热焦炭的初始前驱体[2,3]。因此，这些检测到的碳环含氧有机物极可能是通过 α-羰基醛经羟醛缩合形成的这些初始聚合物的 C—C 键断裂来释放的。图 5-7 显示了以 3-脱氧葡萄糖醛酮为中间体的水热焦炭和含氧碳环有机物的生成路径。可以看到，葡萄糖在水热转化过程中生成的 3-脱氧葡萄糖醛酮可经羟醛缩合、缩醛环化和脱水反应生成含呋喃环结构的聚合物，而这些含呋喃环结构的聚合物可以经水解开环、分子内羟醛缩合反应转化为含碳环结构的聚合物。然后，这些含碳环结构的聚合物可以进一步通过 C—C 键断裂反应释放出多种碳环含氧有机物，包括环己二烯醇衍生物、苯类衍生物和酚类衍生物。

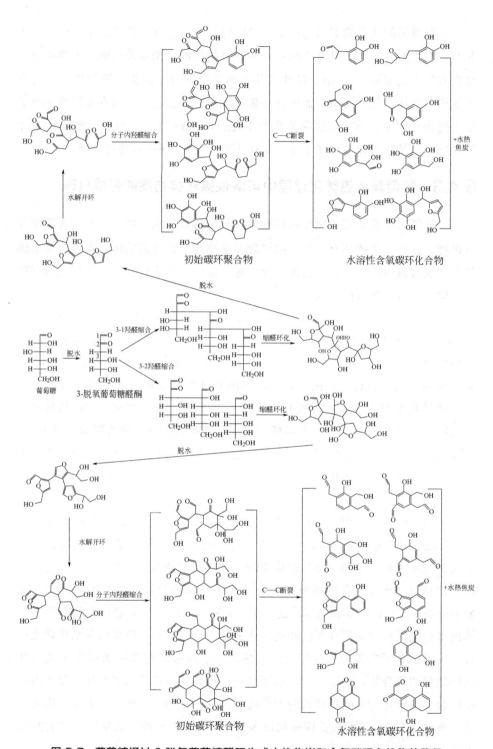

图 5-7 葡萄糖通过 3-脱氧葡萄糖醛酮生成水热焦炭和含氧碳环有机物的路径

5.5 呋喃类化合物水热转化过程中可溶性副产物的鉴定

5.5.1 HMF 水热转化生成的可溶性副产物鉴定

图 5-8 展示了 HMF 水热转化所生成物质的 LC-MS 谱图。在正电离模式下所检测到的物质的分子式包括 $C_5H_8O_3$（12.98min）、$C_{11}H_{12}O_3$（18.09min）、$C_{12}H_{12}O_4$（19.93min）、$C_{12}H_{10}O_5$（21.52min）、$C_{11}H_{14}O_3$（22.56min）、$C_{11}H_{12}O_5$（24.91min）和 $C_{12}H_{12}O_3$（26.37min），而在负电离模式下检测到的物质的分子式包括 $C_6H_6O_3$（10.20min）、$C_5H_8O_3$（12.79min）、$C_{12}H_{10}O_5$（15.48min）、$C_{11}H_{12}O_3$（18.11min）、$C_{12}H_{12}O_4$（19.96min）、$C_{12}H_{10}O_5$（21.50min）、$C_{12}H_{10}O_4$（23.68min）、$C_{11}H_{12}O_5$（24.89min）、$C_{10}H_{12}O_3$（25.60min）和 $C_{12}H_{12}O_3$（26.39min）。很显然，这些检测到的物质包含的碳原子数为 5~12，包含的氧原子数为 2~5。此外，只有 $C_5H_8O_4$、$C_6H_6O_3$、$C_6H_{10}O_4$ 和 $C_5H_8O_3$ 这几种物质的分子量低于 150，而其他物质的分子量在

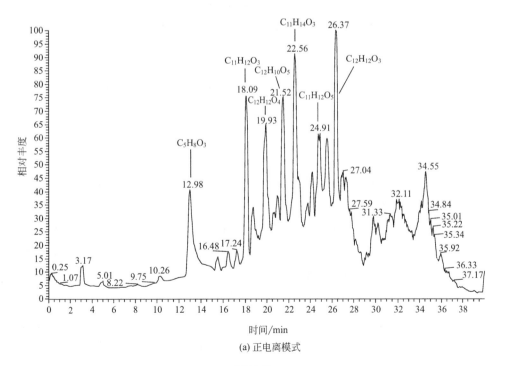

(a) 正电离模式

图 5-8

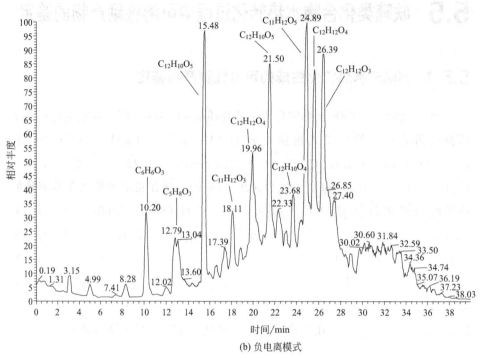

(b) 负电模式

图 5-8　HMF 水热转化所生成物质的 LC-MS 谱图

$166 \sim 234$ 范围内。

根据附表 1 中物质的 MS^2 数据，可以推测出这些物质的核心结构和官能团信息，结果列于表 5-5 中。这些物质的核心结构同样可以分为无环结构、呋喃环结构和碳环结构三类。在这些检测到的物质中，只有物质 $C_5H_8O_3$ (12.98min) 是无环类物质。化合物 $C_6H_6O_3$ (10.20min)、$C_{12}H_{10}O_5$ (15.48min)、$C_{12}H_{10}O_5$ (21.48min) 和 $C_{11}H_{12}O_5$ (24.85min) 都含有呋喃环结构，而化合物 $C_{11}H_{12}O_3$ (18.13min)、$C_{12}H_{12}O_4$ (19.93min)、$C_{11}H_{14}O_2$ (22.56min)、$C_{12}H_{10}O_4$ (23.68min)、$C_{10}H_{14}O_2$ (24.16min)、$C_{12}H_{12}O_4$ (25.60min) 和 $C_{12}H_{12}O_3$ (26.37min) 都是含有碳环结构的物质。很明显，含有碳环结构的物质种类比含有呋喃环结构的物质种类多，表明 HMF 在水热转化过程中同样易生成碳环类物质。绝大多数检测到的物质都含有羟基和羰基，但是只有化合物 $C_5H_8O_3$ (12.98min)、$C_{12}H_{10}O_4$ (23.68min) 和 $C_{12}H_{12}O_4$ (25.60min) 含有羧基。如前所述，$C_5H_8O_3$ (12.98min) 是 HMF 发生水合反应所生成的乙酰丙酸[26]。

表 5-5 HMF 水热转化鉴定的可溶性化合物

时间/min	分子式	分子量	核心结构	官能团	DBE	DBE/C	$(H-2O)/C$
10.20	$C_6H_6O_3$	126.03	呋喃环	羟基、羰基	4	0.67	0
12.98	$C_5H_8O_3$	116.05	链式醛/酸	羰基、羧基	2	0.4	0.4
15.48	$C_{12}H_{10}O_5$	234.05	呋喃环	羟基、羰基	8	0.67	0
18.07~18.13	$C_{11}H_{12}O_3$	192.08	六元碳环	羟基、羰基	6	0.55	0.55
19.93~19.96	$C_{12}H_{12}O_4$	220.07	六元碳环	羟基、羰基	7	0.58	0.33
21.48~21.50	$C_{12}H_{10}O_5$	234.05	呋喃环	羟基、羰基	8	0.67	0
22.56	$C_{11}H_{14}O_3$	194.09	六元碳环	羟基、羰基	5	0.45	0.73
23.68	$C_{12}H_{10}O_4$	218.06	呋喃环	羟基、羰基	8	0.67	0.17
24.16	$C_{10}H_{14}O_2$	166.10	六元碳环	羟基、羰基	4	0.4	1
24.75~24.85	$C_{11}H_{12}O_5$	224.07	呋喃环	羟基、羰基	6	0.55	0.18
25.49~25.60	$C_{10}H_{12}O_3$	220.07	六元碳环	羟基、羰基、羧基	5	0.5	0.6
26.37	$C_{12}H_{12}O_3$	204.08	六元碳环	羟基、羰基	7	0.58	0.5

通过这些物质的分子式可以计算出这些物质的 DBE、DBE/C 和 $(H-2O)/2$ 的值,这些数据同样列于表 5-5 中。DBE/C 表示每个碳原子所分摊的当量双键量,在水热条件下,这个值在水加成反应时会下降,在脱水反应时则会上升。HMF 的 DBE/C 值为 0.67,而这些检测到的化合物的 DBE/C 值都低于 0.67,表明这些化合物的形成都涉及了 HMF 的水加成反应。这一结果进一步表明,水解开环反应是 HMF 在水热转化过程中的关键步骤。

如上所述,$(H-2O)/C$ 的值表示每个碳原子所分摊的净氢原子数,这个值可以用于判断这些物质生成过程中是否发生了 C—C 键水解断裂反应[27]。HMF 的 $(H-2O)/C$ 为 0,而在检测到的物质中,除了 $C_6H_6O_3$ (10.20min)、$C_{12}H_{10}O_5$ (15.48min) 和 $C_{12}H_{10}O_5$ (21.48min) 之外,其他检测到的物质的 $(H-2O)/C$ 都不为 0,表明 HMF 生成这些物质时都发生了 C—C 键水解断裂反应。

根据这些检测到的物质的核心结构和官能团信息,可以推测这些化合物可能的分子结构,结果列于图 5-9 中。可以看到,HMF 水热转化的产物分布与葡萄糖水热转化生成的产物种类相似,表明 HMF 确实是葡萄糖降解过程中的关键中间产物。

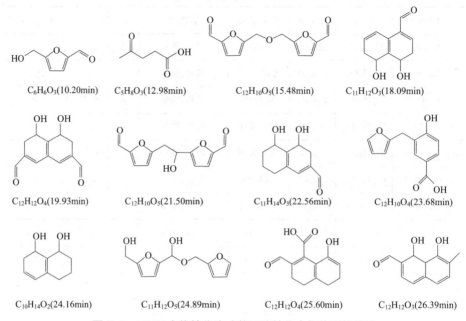

图 5-9 HMF 水热转化生成的可溶性化合物的可能结构

5.5.2 糠醛水热转化生成的可溶性物质鉴定

糠醛水热转化生成的可溶性物质的 LC-MS 谱图及主要峰及所对应物质的分子式（根据附表 2 中的 MS 数据）如图 5-10 所示。在正电离模式下检测到的物质分子式为 $C_{10}H_8O_4$（20.96min）、$C_{15}H_{12}O_5$（21.52min）、$C_{10}H_{10}O_3$（22.21min）、$C_{10}H_8O_4$（24.15min）、$C_{11}H_{10}O_5$（25.57min）和 $C_{15}H_{12}O_5$（26.38min），而在负电离模式下检测到的物质的分子式为 $C_5H_8O_4$（10.13min）、$C_4H_6O_4$（10.57min）、$C_8H_{10}O_3$（16.60min）、$C_{10}H_8O_4$（17.01min）、$C_{12}H_{14}O_4$（18.39min）、$C_9H_{12}O_3$（19.12min）、$C_9H_{12}O_4$（19.80min）、$C_{10}H_8O_4$（20.94min）、$C_{15}H_{12}O_5$（21.50min）、$C_{10}H_8O_4$（24.13min）、$C_{10}H_8O_4$（25.22min）和 $C_{15}H_{12}O_5$（26.28min）。很显然，除了 $C_5H_8O_4$（10.13min）和 $C_4H_6O_4$（10.57min）这两种物质外，其他物质含有的碳原子数为 8~15，而氧原子数为 3~5。

类似地，通过对这些物质的 MS^2 数据（见附表 2）的分析，可以确定这些物质的核心结构和官能团（表 5-6）。在所有检测到的物质中，只有物质 $C_5H_8O_4$（10.13min）和 $C_4H_6O_4$（10.57min）是无环结构且分子量低于 140，而其他检测到的物质都是分子量为 154~272 的碳环类物质。羟基和羰基这

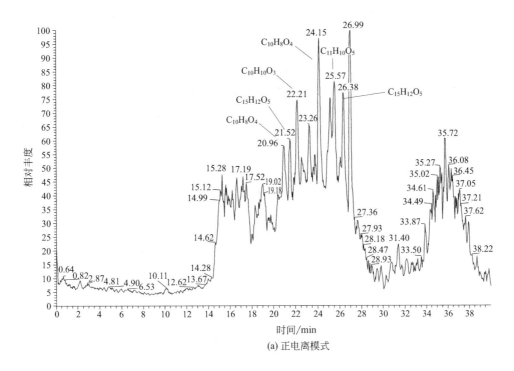

(a) 正电离模式

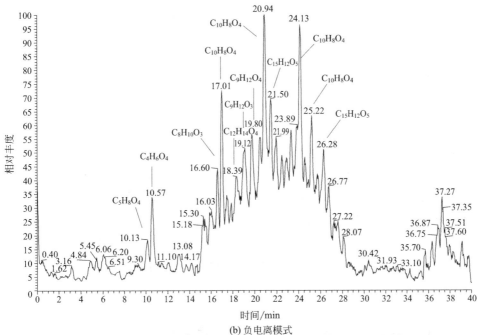

(b) 负电离模式

图 5-10　糠醛水热转化过程中所生成物质的 LC-MS 谱图及所对应物质的分子式

两种官能团都被证明存在于这些检测到的物质中，但是羧基只存在于 $C_5H_8O_4$(10.13min)、$C_4H_6O_4$(10.57min)和$C_{15}H_{12}O_5$(26.38min)这三种物质中。

表 5-6 糠醛水热转化鉴定的可溶性化合物

时间/min	分子式	分子量	核心结构	官能团	DBE	DBE/C	$(H-2O)/C$
10.13	$C_5H_8O_4$	132.04	链式醛/酸	羟基、羰基、羧基	2	0.4	0
10.57	$C_4H_6O_4$	118.02	链式醛/酸	羟基、羰基、羧基	2	0.5	-0.5
16.60	$C_8H_{10}O_3$	154.06	六元碳环	羟基、羰基	4	0.5	0.5
17.01	$C_{10}H_8O_3$	192.04	六元碳环	羟基、羰基	7	0.7	0
18.39	$C_{12}H_{14}O_4$	222.09	六元碳环	羟基、羰基	6	0.5	0.5
19.80	$C_8H_{12}O_4$	184.07	六元碳环	羟基、羰基	4	0.44	0.44
20.94	$C_{11}H_{12}O_3$	192.04	六元碳环	羟基、羰基	6	0.55	0.55
21.50	$C_{15}H_{12}O_5$	272.07	六元碳环	羟基、羰基	10	0.67	0.13
22.21	$C_{10}H_{10}O_3$	178.06	六元碳环	羟基、羰基	6	0.6	0.4
23.26	$C_{10}H_8O_4$	192.04	六元碳环	羟基、羰基	7	0.7	0
24.13	$C_{10}H_8O_4$	192.04	六元碳环	羟基、羰基	7	0.7	0
25.22	$C_{10}H_8O_4$	192.04	六元碳环	羟基、羰基	7	0.7	0
25.57	$C_{11}H_{10}O_5$	222.05	六元碳环	羟基、羰基	7	0.64	0
26.28~26.38	$C_{15}H_{12}O_5$	272.07	六元碳环	羟基、羰基、羧基	10	0.67	0.13

糠醛的 DBE/C 为 0.8，而这些检测到的物质的 DBE/C 都低于 0.8，表明从糠醛生成这些物质的过程中同样涉及水解开环反应。糠醛的 $(H-2O)/C$ 是 0，而检测到的所有物质中只有分子式为 $C_{10}H_8O_4$ 的物质的 $(H-2O)/C$ 是 0，表明形成 $C_{10}H_8O_4$ 的过程中不涉及 C—C 键断裂反应，而形成其他物质的过程都涉及 C—C 键断裂反应。

图 5-11 展示了这些检测到的物质可能的分子结构。所有的这些碳环类物质都含有双环结构，这是因为一些含有双环结构的离子碎片如[C_9H_9]、[$C_{10}H_{11}$]、[C_9H_9O]和[$C_{10}H_{11}O$]等存在于这些物质的 MS^2 质谱图中。

5.5.3 糠醇水热转化生成的可溶性物质鉴定

图 5-12 展示了糠醇水热转化过程中生成的物质的 LC-MS 谱图。在正电离模

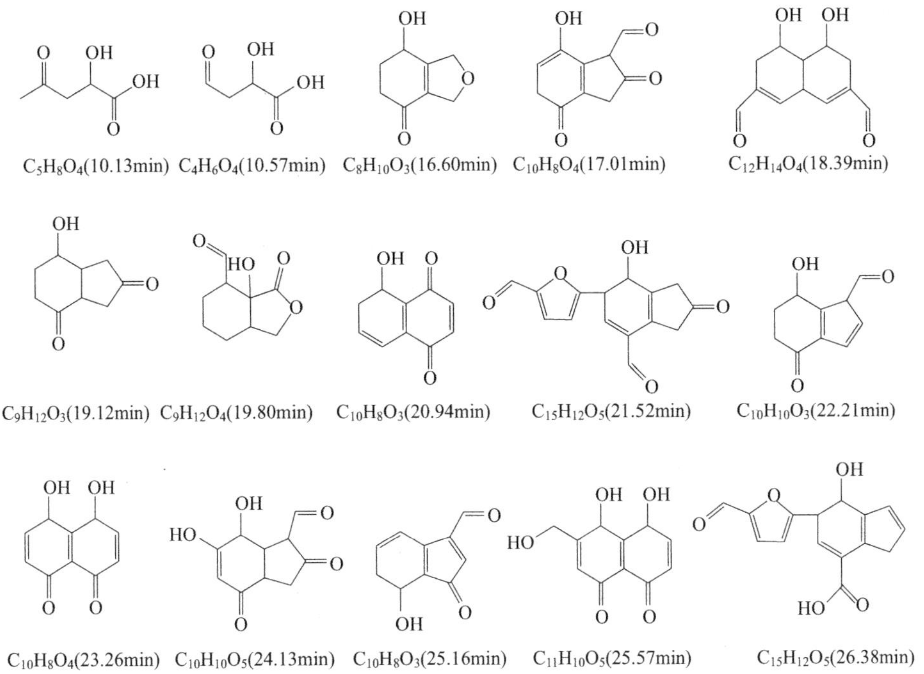

$C_5H_8O_4$(10.13min)　$C_4H_6O_4$(10.57min)　$C_8H_{10}O_3$(16.60min)　$C_{10}H_8O_4$(17.01min)　$C_{12}H_{14}O_4$(18.39min)

$C_9H_{12}O_3$(19.12min)　$C_9H_{12}O_4$(19.80min)　$C_{10}H_8O_3$(20.94min)　$C_{15}H_{12}O_5$(21.52min)　$C_{10}H_{10}O_3$(22.21min)

$C_{10}H_8O_4$(23.26min)　$C_{10}H_{10}O_5$(24.13min)　$C_{10}H_8O_3$(25.16min)　$C_{11}H_{10}O_5$(25.57min)　$C_{15}H_{12}O_5$(26.38min)

图 5-11　糠醛水热转化过程中鉴定的可溶性化合物

式下鉴定出的物质的分子式为 $C_{10}H_{12}O_4$(17.18min)、$C_9H_{14}O_3$(18.71min)、$C_{10}H_{12}O_4$(19.40min)、$C_{10}H_{10}O_3$(19.88min)、$C_{11}H_{12}O_3$(21.42min)、$C_{14}H_{12}O_2$(22.87min)、$C_{10}H_{10}O_3$(26.37min)、$C_{11}H_{12}O_3$(27.29min) 和 $C_{14}H_{16}O_3$(27.86min),而在负电离模式下鉴定出的物质的分子式为 $C_4H_6O_4$(10.53min)、$C_5H_8O_3$(12.99min)、$C_5H_6O_2$(13.45min)、$C_8H_{10}O_3$(15.46min)、$C_{10}H_{12}O_4$(16.03min)、$C_{10}H_{12}O_4$(17.20min)、$C_{10}H_{12}O_4$(18.01min)、$C_{10}H_{10}O_3$(19.98min) 和 $C_{15}H_{14}O_4$(21.12min)。很显然,$C_5H_8O_3$ 是糠醇水合重排所生成的乙酰丙酸[25],而 $C_5H_6O_2$ 是未转化的糠醇。除了 $C_4H_6O_4$(10.53min)、$C_5H_8O_3$(12.99min) 和 C_5H_6O(13.45min),所有其他鉴定出的物质都含有 8~15 个碳原子和 2~4 个氧原子。$C_{10}H_{10}O_3$(19.98min) 在正电离和负电离模式下都是最主要的产物。

同样地,根据这些物质的 MS^2 数据（见附表3）,可以确定这些物质的核心结构和官能团种类和数量,进而推测出这些物质可能的分子结构。糠醇生成的物质的核心结构和官能团信息如表 5-7 所示。在这些检测到的物质中,只有 $C_4H_6O_4$(10.53min)、$C_5H_8O_3$(12.99min) 和 $C_5H_6O_2$(13.45min) 不含碳环结构,而其他物质都是含有羟基和羰基的碳环类物质。这些物质的碳链都比糠

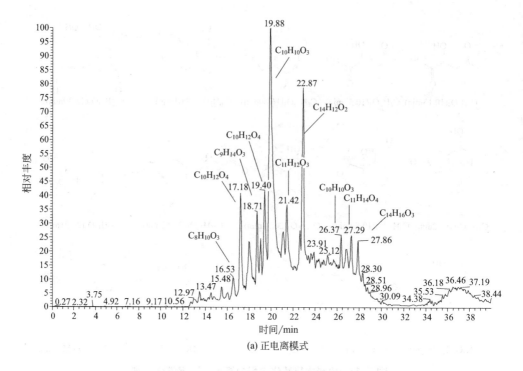

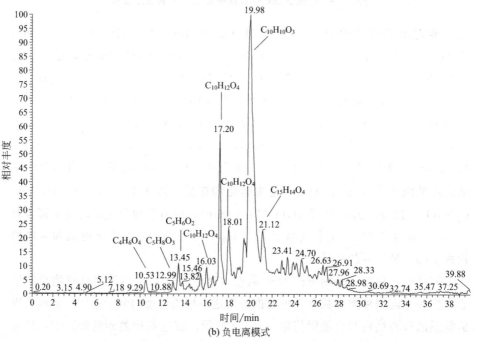

图 5-12 糠醇水热转化过程中所生成物质的 LC-MS 谱图

醇的碳链长，表明从糠醇生成这些物质的过程中涉及碳链增长反应，比如羟醛缩合反应。

糠醇的 DBE/C 是 0.6，而绝大多数检测到的可溶性物质的 DBE/C 都低于 0.6，同样表明从糠醇生成这些物质的过程中涉及水解开环反应。糠醇的 $(H-2O)/C$ 是 0.4，而绝大多数检测到的物质的 $(H-2O)/C$ 都不等于 0.4，表明这些物质的生成过程中涉及了 C—C 键水解断裂反应。$C_{10}H_{10}O_3$（19.98min）这种最主要产物的 $(H-2O)/C$ 的值为 0.4，表明从糠醇生成这种物质的过程不涉及 C—C 键水解断裂反应。

表 5-7 糠醇水热转化鉴定的可溶性化合物

时间/min	分子式	分子量	核心结构	官能团	DBE	DBE/C	$(H-2O)/C$
10.53	$C_4H_6O_4$	118.03	链式醛/酸	羟基、羰基、羧基	2	0.5	-0.5
12.99	$C_5H_8O_3$	116.05	链式醛/酸	羰基、羧基	2	0.4	0.4
13.45	$C_5H_6O_2$	98.04	呋喃环	羟基	3	0.6	0.4
15.46	$C_8H_{10}O_3$	154.06	六元碳环	羟基、羰基	4	0.5	0.5
16.03	$C_{10}H_{12}O_4$	196.07	六元碳环	羟基、羰基、羧基	5	0.5	0.4
16.53	$C_8H_{10}O_3$	154.06	六元碳环	羟基、羰基	4	0.5	0.5
17.20	$C_{10}H_{12}O_4$	196.07	六元碳环	羟基、羰基、羧基	5	0.5	0.4
18.01	$C_{10}H_{12}O_4$	196.07	六元碳环	羟基、羰基、羧基	5	0.5	0.4
18.71	$C_8H_{14}O_3$	170.09	六元碳环	羟基、羰基	3	0.33	0.89
19.40	$C_{10}H_{12}O_4$	196.07	六元碳环	羟基、羰基	5	0.5	0.4
19.98	$C_{10}H_{10}O_3$	178.06	六元碳环	羟基、羰基	6	0.6	0.4
21.12	$C_{15}H_{14}O_4$	258.09	六元碳环	羟基、羰基	9	0.6	0.4
21.42	$C_{11}H_{12}O_4$	192.08	六元碳环	羟基、羰基、羧基	6	0.55	0.55
22.87	$C_{14}H_{12}O_2$	212.08	六元碳环	羟基、羰基	9	0.64	0.57
26.37	$C_{10}H_{10}O_3$	178.06	六元碳环	羟基、羰基	6	0.6	0.4
27.29	$C_{11}H_{12}O_3$	212.08	六元碳环	羟基、羰基	6	0.55	0.55
27.86	$C_{14}H_{16}O_3$	232.11	六元碳环	羟基、羰基	7	0.5	0.71

图 5-13 展示了这些物质可能的分子结构。可以看到，与 HMF 和糠醛在水热转化过程中生成的可溶性物质相似，糠醇在水热转化过程中生成了大量含有羟基和羰基的双碳环类物质。

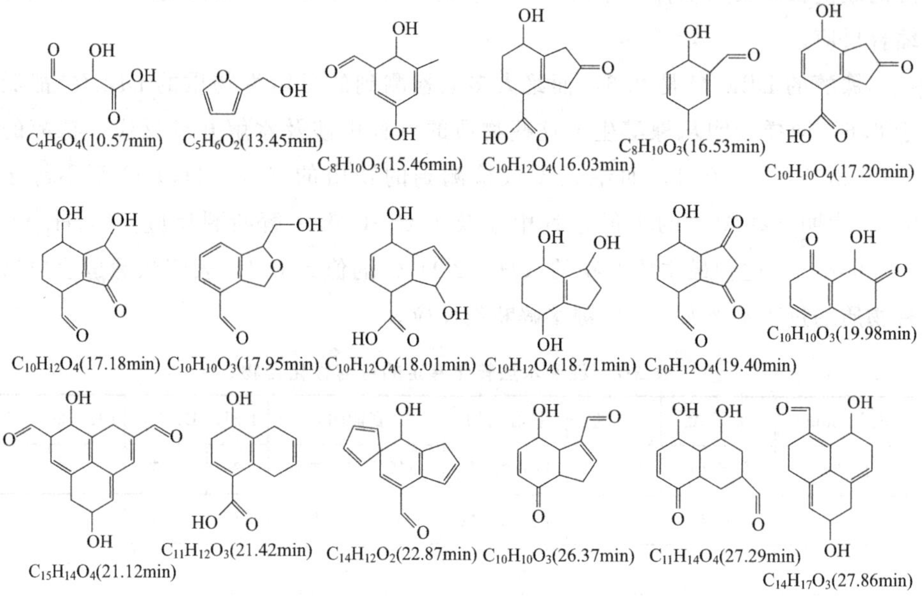

图 5-13 糠醇水热转化过程中鉴定的可溶性化合物

5.5.4 呋喃类物质的水热转化路径

通过对呋喃类物质水热转化所生成的可溶性物质进行鉴定，可以发现，这些呋喃类物质在水热转化过程中都易生成大量含有羟基和羰基的碳环类物质。绝大多数检测到的碳环类物质的 DBE/C 都低于这些呋喃类物质，表明在生成这些碳环类物质的过程中都涉及了水解开环反应。这可能是呋喃类物质只可在水中生成水热焦炭和碳环类物质的原因。第四章的研究表明碳环类物质可以通过水解开环反应形成链式多羰基化合物，而这些链式多羰基化合物则易发生羟醛缩合反应生成水热焦炭[1,2,26,28]。所以，呋喃类物质水解开环所生成的链式多羰基醛是生成水热焦炭和碳环类物质的关键中间体。另一方面，这些检测到的碳环类物质的碳链都比呋喃类物质的碳链长，表明这些碳环类物质的生成涉及了碳链增长反应。由于呋喃类物质通过水解开环反应所生成的链式多羰基化合物易于发生羟醛缩合反应[2,28,29]，所以，羟醛缩合反应是呋喃类物质生成碳环类物质的关键步骤。

综上所述，在呋喃类物质水热转化生成碳环类物质的过程中都涉及了水解开环、碳链增长和形成碳环等过程。据此，我们分析了 HMF、糠醛和糠醇在水热条件下转化为碳环类物质的路径。

如图 5-14 所示，HMF 经水解开环反应可生成 2,5-二氧代-6-羟基己醛[3,28,29]。这种链式多羰基化合物含有 3 个羰基和 1 个羟基，因此它能够通过分子间的羟醛缩合反应生成初始的链状聚合物。这些链状聚合物同样含有大量的羟基和羰基，因此它们可以发生分子内羟醛缩合反应而生成含有碳环结构且富含羟基和羰基的化合物。然后，这些含有碳环结构的化合物可以继续发生分子内羟醛缩合反应而生成含有双环结构的物质，或者发生 C—C 键水解断裂反应而生成检测到的可溶性碳环类物质。

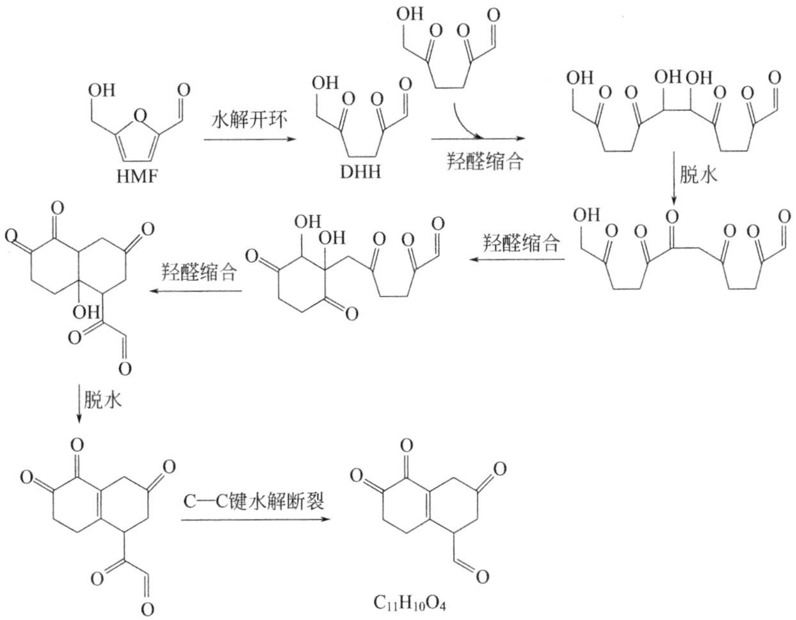

图 5-14 由 HMF 生成碳环化合物 $C_{11}H_{10}O_4$ 的可能路径

糠醛和糠醇也能够通过相似的路径而生成碳环类物质。上面的研究表明糠醛在水热转化过程中易生成含有羟基和羰基的碳环类物质 $C_{10}H_8O_4$，图 5-15 展示了糠醛生成 $C_{10}H_8O_4$ 的路径。可以看到，糠醛通过水解开环所生成的链式 α-羰基醛 2-氧代戊二醛同样可以通过羟醛缩合生成含有大量羰基的链式二聚体，该二聚体则能够通过分子内羟醛缩合和脱水反应生成含有羟基和羰基的双碳环化合物 $C_{10}H_8O_4$。

同理，图 5-16 展示了糠醇生成 $C_{10}H_{10}O_3$ 的路径。可以看到，糠醇生成该物质的过程涉及糠醇水解开环生成链式多羰基化合物，链式多羰基化合物发生羟醛缩合反应生成长链多羰基化合物，长链多羰基化合物发生分子内羟醛缩合反应和脱水反应生成含有双碳环结构的 $C_{10}H_{10}O_3$。糠醛和糠醇在水热转化

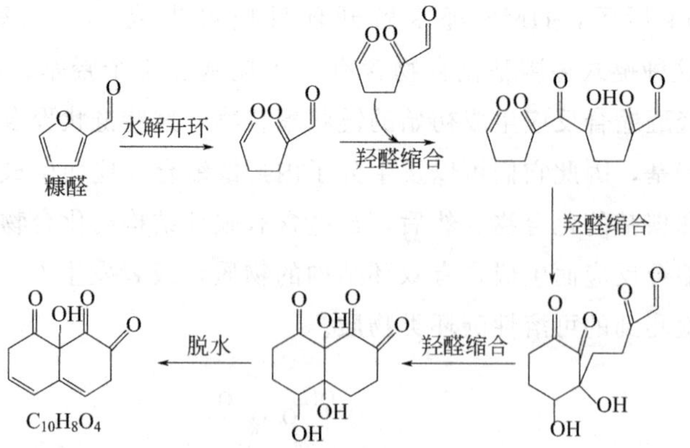

图 5-15 糠醛生成 $C_{10}H_8O_4$ 的路径

过程中发生 C—C 键水解断裂的反应产物较少，表明这两种物质在水热转化过程中生成的物质不易发生 C—C 键水解断裂反应。

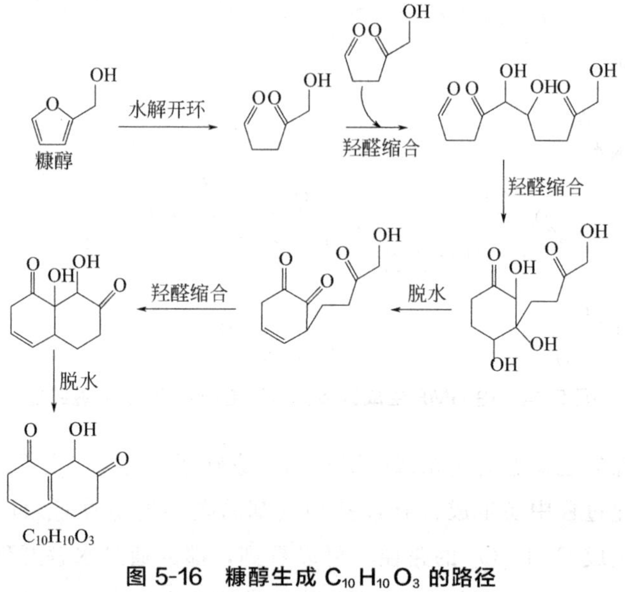

图 5-16 糠醇生成 $C_{10}H_{10}O_3$ 的路径

综上所述，从呋喃类物质生成碳环类物质包括下述反应步骤：①呋喃环结构发生水解开环生成链式多羰基化合物；②生成的链式多羰基化合物继续发生分子间羟醛缩合反应而生成富含羟基和羰基的链式初始聚合物；③链式初始聚合物发生分子内羟醛缩合反应，生成含有六元碳环结构的聚合物；④碳环结构的聚合物发生 C—C 键水解断裂反应而生成含有羟基和羰基的水溶性碳环类物质。

5.6 C—C 键水解断裂反应

上述研究结果表明，在葡萄糖、HMF、糠醛和糠醇的水热转化过程中，会很频繁地发生某种由于水加成导致的 C—C 键断裂反应。这种 C—C 键水解断裂反应是一种分子内氧化还原反应，会释放出一种氧化产物和一种还原产物。其中，水分子中的氢氧根和氢原子分别加成到被氧化和被还原的基团上，生成 $(H-2O)/C$ 值低于反应物的氧化产物和 $(H-2O)/C$ 值高于反应物的还原产物（图 5-17）。葡萄糖、果糖、木糖等碳水化合物都是 $(H-2O)/C$ 值为 0 的物质，而它们在水热转化过程中生成了部分 $(H-2O)/C$ 大于 0 的物质和部分 $(H-2O)/C$ 小于 0 的物质，表明这些物质的生成过程中发生了 C—C 键水解断裂反应。

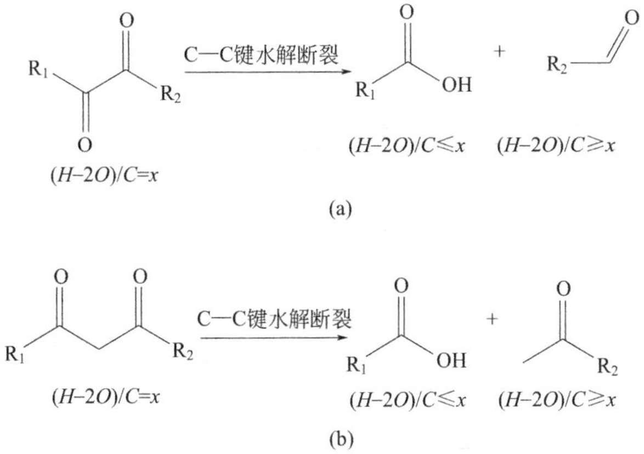

图 5-17 α-二羰基结构（a）和 β-二羰基结构（b）物质的 C—C 键水解断裂反应

逆羟醛缩合反应是一种会导致 C—C 键断裂的反应，但是对于 $(H-2O)/C$ 为 0 的物质比如葡萄糖、果糖、木糖等，它们通过逆羟醛缩合反应所生成的产物的 $(H-2O)/C$ 仍然为 0。比如，葡萄糖的 $(H-2O)/C$ 为 0，它发生逆羟醛缩合反应能够生成乙醇醛 $C_2H_4O_2$、甘油醛 $C_3H_6O_3$、二羟基丙酮 $C_3H_6O_3$ 和赤藓糖 $C_4H_8O_4$[30,31]，而这些物质的 $(H-2O)/C$ 也都为 0。因此，笔者认为，上述模型化合物在水热转化过程中发生了不同于逆羟醛缩合的 C—C 键水解断裂反应。

因为 C—C 键水解断裂反应可以生成可溶性物质，所以掌握 C—C 键水解断裂反应的原理及其发生的位点，从而寻找能够促进 C—C 键水解断裂的催化剂，对生物质水热催化转化具有重要意义。C—C 键水解断裂反应应当发生在具有某些特殊官能团的化合物上。有研究发现，含有二羰基结构的物质可以在水热环境中通过 C—C 键水解断裂生成羧酸，并释放出相应的醛/酮[32-34]。比如，HMF 生成乙酰丙酸和甲酸的过程涉及中间体 2,5-二氧代-3-己烯醛这种 α-羰基醛发生 C—C 键水解断裂反应，生成一个甲酸分子和一个 4-氧代-2-戊烯醛[25,35]，而葡萄糖水热转化生成乙酸的过程涉及 β-二羰基化合物的 C—C 键水解反应[36-39]。此外，Davidek 等发现 β-二羰基化合物 2,4-戊二酮可以在水热转化过程中生成乙酸和丙酮，而 α-二羰基化合物 2,3-戊二酮在水热氧化过程中可以生成乙酸和丙酸[40,41]。另一方面，前面的研究表明，HMF 和糠醛在水解开环后可以分别生成 2,5-二氧代-6-羟基己醛和 2-氧代戊二醛这两种含有 α-二羰基结构的化合物[3,26,28]，而这些链式多羰基化合物在聚合后同样能够生成含有 α-二羰基结构和 β-二羰基结构的初始聚合物。因此，笔者认为，上述模型化合物在水热转化过程中的 C—C 键水解断裂反应发生在含有二羰基结构的物质上。

5.7　α-羰基醛的生成与转化路径

上述研究表明碳水化合物在水热转化过程中所生成的 α-羰基醛是碳水化合物生成羧酸、呋喃类物质、碳环类物质和水热焦炭的关键中间体。因此，分析 α-羰基醛的生成和转化路径，对理解碳水化合物的水热转化过程具有重要意义。

如图 5-18 所示，碳水化合物在水热转化过程中可以生成多种 α-羰基醛。以葡萄糖为例，它在水热转化过程中可以通过三种途径生成 α-羰基醛[21,26,31]：①葡萄糖可以通过缩醛环化和脱水反应生成 HMF，而 HMF 可以通过水解开环反应而生成 α-羰基醛，如 2,5-二氧代-6-羟基己醛、2,5-二氧代-3-己烯醛[26,28,29,42]；②葡萄糖可以通过 β-消除和酮式-烯醇互变异构而生成 α-羰基醛（3-脱氧葡萄糖醛酮）[21]；③葡萄糖可以通过逆羟醛缩合反应而生成甘油醛、1,3-二羟基丙酮和赤藓糖，而这些物质可以发生 β-消除反应生成 α-羰基醛，如丙酮醛和 3-脱氧赤藓酮醛[2,43,44]。

生成的 α-羰基醛可以通过下述路径被转化：①α-羰基醛可以发生坎尼扎罗反应而形成 α-羟基酸[30,31]；②α-羰基醛可以发生羟醛缩合反应而生成水热焦炭并副产碳环类物质[3,26,28]；③α-羰基醛可以发生 C—C 键水解断裂而生成短链羧酸和醛[40,41]。由于 α-羰基醛发生坎尼扎罗反应和 C—C 键水解断裂反应

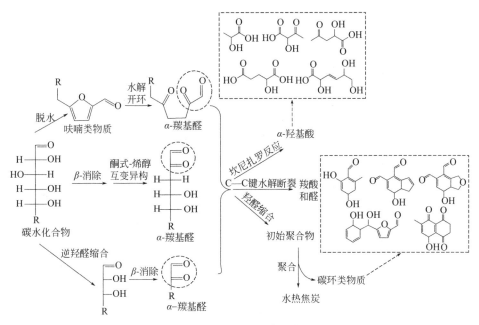

图 5-18 α-羰基醛的生成与转化路径

都能够生成可溶性物质而避免生成水热焦炭,因此,笔者认为,深入研究对坎尼扎罗反应和 C—C 键水解断裂反应具有较好催化作用的催化剂,是碳水化合物在水热转化过程中抑制水热焦炭生成的关键。

5.8 碳水化合物的降解路径

上述研究表明,α-羰基醛可以发生坎尼扎罗反应而生成 α-羟基酸,发生 C—C 键水解断裂反应而生成醛和酸,或者经过聚合反应而生成水热焦炭和碳环类物质。认识 α-羰基醛的降解路径有助于分析碳水化合物的降解路径,因为醛糖(如葡萄糖和木糖)很容易通过 β-消除和酮式-烯醇互变异构反应而生成 α-羰基醛[21,45-48]。根据 α-羰基醛的降解路径,并结合碳水化合物生成 α-羰基醛的路径,笔者对葡萄糖和木糖的水热降解路径进行了分析。

(1)葡萄糖和果糖的降解路径

图 5-19 为葡萄糖和果糖的水热降解路径。可以看到,链式葡萄糖在水热条件下经 β-消除和酮式-烯醇互变异构得到 3-脱氧葡萄糖醛酮,这是一种 α-羰基醛,它会经由下述四种路径被转化[21,49,50]:①3-脱氧葡萄糖醛酮经坎尼扎罗反应生成一种 α-羟基酸,即 2,4,5,6-四羟基己酸[21,51];②3-脱氧葡萄糖醛

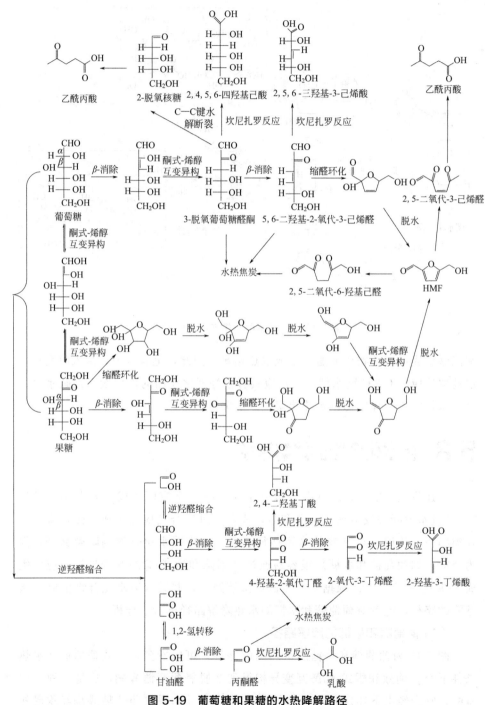

图 5-19 葡萄糖和果糖的水热降解路径

酮经由缩醛环化和脱水反应生成 HMF；③3-脱氧葡萄糖醛酮发生 C—C 键水解断裂反应生成 2-脱氧核糖，进而生成糠醇和乙酰丙酸；④3-脱氧葡萄糖醛酮分子间发生羟醛缩合、缩醛环化和脱水反应生成水热焦炭。可以看出，由于 3-脱氧葡萄糖醛酮的降解路径复杂，导致葡萄糖在水热降解过程中的路径比较复杂，HMF 等平台化学品的选择性低。

不同于葡萄糖，果糖在水溶液中呈呋喃环结构，所以它可以直接脱水得到 HMF[33]。此外，果糖发生 β-消除反应得到的是 α-羰基酮而不是 α-羰基醛，故果糖在水热降解时生成 HMF 的选择性较高。

如上所述，葡萄糖都能够在水热条件下生成 HMF。HMF 在水热条件下并不稳定，它易发生水解开环继续生成两种新的 α-羰基醛，即 2,5-二羰基-6-羟基己醛和 2,5-二氧代-3-己烯醛[28,29]。如图 4-3 所示，这两种 α-羰基醛同样可以经过多种路径被转化，从而生成乙酰丙酸和水热焦炭。

另一方面，葡萄糖和果糖都可以发生逆羟醛缩合反应形成碳链较短的多羟基醛（酮），并继续反应生成其他衍生化合物。葡萄糖通过逆羟醛缩合反应生成乙醇醛和赤藓糖[40]，而赤藓糖也可以通过 β-消除和酮式-烯醇互变异构反应而生成相应的 α-羰基醛，即 4-羟基-2-氧代丁醛[40]。4-羟基-2-氧代丁醛可以经由三种路径被转化[52-54]：①4-羟基-2-氧代丁醛发生坎尼扎罗反应生成 2,4-二羟基丁酸[45]；②4-羟基-2-氧代丁醛继续发生 β-消除和坎尼扎罗反应生成 2-氧代-3-丁烯醛，而 2-氧代-3-丁烯醛则发生坎尼扎罗反应生成 2-羟基-3-丁烯酸[45,54]；③4-羟基-4-氧代丁醛经由羟醛缩合反应、缩醛环化和脱水反应生成水热焦炭。果糖发生逆羟醛缩合反应生成两种丙糖，即二羟基丙酮和甘油醛[55,56]。二羟基丙酮能够通过 1,2-氢转移生成甘油醛，而甘油醛则发生 β-消除和酮式-烯醇互变异构生成丙酮醛，丙酮醛也是一种 α-羰基醛，它可经由坎尼扎罗反应生成乳酸[55,56]，或者经由分子内羟醛缩合和缩醛反应生成水热焦炭。

（2）木糖的水热降解路径

图 5-20 展示了木糖的水热降解路径。与葡萄糖类似，链式木糖在水热条件下可以发生 β-消除和酮式-烯醇互变异构反应而生成链式的 3-脱氧木糖醛酮，而 3-脱氧木糖醛酮则通过五种途径被转化：①3-脱氧木糖醛酮经由坎尼扎罗反应而生成 2,4,5-三羟基戊酸；②3-脱氧木糖醛酮经过缩醛环化和脱水反应而生成糠醛；③3-脱氧木糖醛酮发生 C—C 键水解断裂而生成 2-脱氧赤藓糖，并经缩醛环化和脱水反应生成呋喃；④3-脱氧木糖醛酮通过羟醛缩合、缩醛环化和脱水反应而生成水热焦炭；⑤3-脱氧木糖醛酮发生 β-消除反

应而生成 5-羟基-2-氧代-3-戊烯醛，并进而通过坎尼扎罗反应生成 2,5-二羟基-3-戊烯酸。木糖生成的糠醛同样可以发生水解开环反应而生成 2-氧代-戊二醛，而 2-氧代-戊二醛同样是 α-羰基醛，可以继续通过羟醛缩合、缩醛环化和脱水反应而生成水热焦炭，或者通过坎尼扎罗反应而生成新的 α-羟基酸。

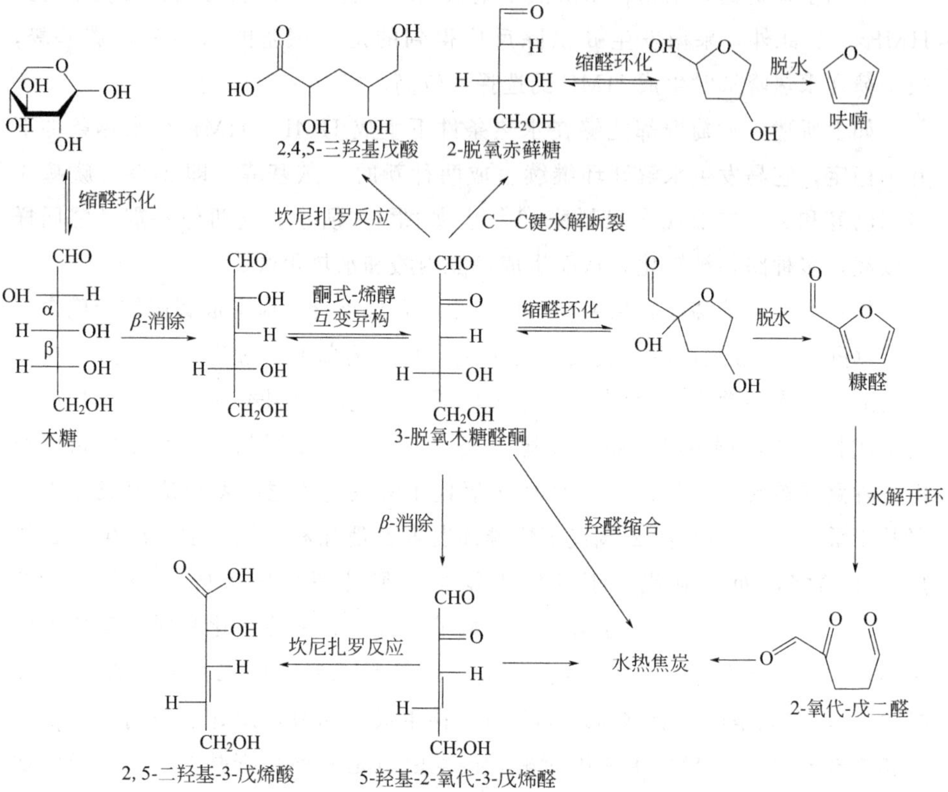

图 5-20　木糖的水热降解路径

可以看到，由于 α-羰基醛降解路径的复杂性，导致碳水化合物的水热降解路径极其复杂。研究上述各种基本反应，对于碳水化合物高选择性转化有较为重要的意义。

5.9　结论

利用 LC-MS 和 LC-MS^2 对葡萄糖、HMF、糠醛和糠醇在水热转化过程中形成的水溶性副产物进行了分析。在水热转化过程中，上述模型化合物时会生成多种分子量在150—260之间的含羟基和羰基的呋喃类和碳环类含氧有机物。这些碳环类物质的形成涉及呋喃环的水解开环反应、分子间羟醛缩合反应、分子内羟醛缩合反应和C—C键水解反应。C—C键水解断裂反应被认为发生在含有α-二羰基结构和β-二羰基结构的化合物上。进一步研究C—C键水解断裂的机理及开发对此反应具有较好催化作用的催化剂,是碳水化合物水热转化过程中抑制水热焦炭生成的一个重要方向。碳环含氧有机物的形成间接证明了水热焦炭是由α-羰基醛的羟醛缩合形成的。

参 考 文 献

[1] Shi N, Liu Q, Cen H, Ju R, He X, Ma L. Formation of humins during degradation of carbohydrates and furfural derivatives in various solvents. Biomass Conversion and Biorefinery, 2020, 10 (2): 277-287.

[2] Shi N, Liu Q Y, Ju R M, He X, Zhang Y L, Tang S Y, Ma L L. Condensation of α-carbonyl aldehydes leads to the formation of solid humins during the hydrothermal degradation of carbohydrates. Acs Omega, 2019, 4 (4): 7330-7343.

[3] Shi N, Liu Q, He X, Wang G, Chen N, Peng J, Ma L. Molecular structure and formation mechanism of hydrochar from hydrothermal carbonization of carbohydrates. Energy and Fuels, 2019, 33 (10): 9904-9915.

[4] Shi N, Liu Q Y, He X, Cen H, Ju R M, Zhang Y L, Ma L L. Production of lactic acid from cellulose catalyzed by easily prepared solid $Al_2(WO_4)_3$. Bioresource Technology Reports, 2019, 5: 66-73.

[5] Shi N, Liu Q Y, Wang T J, Zhang Q, Tu J L, Ma L L. Conversion of cellulose to 5-hydroxymethylfurfural in water-tetrahydrofuran and byproducts identification. Chinese Journal of Chemical Physics, 2014, 27 (6): 711-717.

[6] Shi N, Liu Q Y, Zhang Q, Wang T J, Ma L L. High yield production of 5-hydroxymethylfurfural from cellulose by high concentration of sulfates in biphasic system. Green Chemistry, 2013, 15 (7): 1967-1974.

[7] Maruani V, Narayanin-Richenapin S, Framery E, Andrioletti B. Acidic hydrothermal dehydration of D-glucose into humins: identification and characterization of intermediates. Acs Sustainable Chemistry and Engineering, 2018, 6 (10): 13487-13493.

[8] Buendia-Kandia F, Mauviel G, Guedon E, Rondags E, Petitjean D, Dufour A. Decomposition of cellulose in hot-compressed water: detailed analysis of the products and effect of operating conditions. Energy and Fuels, 2017, 32 (4): 4127-4138.

[9] Gagić T, Perva-Uzunalić A, Knez Ž, Škerget M. Hydrothermal degradation of cellulose at temperature from 200 to 300℃. Industrial and Engineering Chemistry Research, 2018, 57 (18): 6576-6584.

[10] Poerschmann J, Weiner B, Koehler R, Kopinke F D. Hydrothermal carbonization of glucose, fructose, and xylose—identification of organic products with medium molecular masses. Acs Sustainable Chemistry and Engineering, 2017, 5 (8): 6420-6428.

[11] Shi N, Liu Q Y, Wang T J, Zhang Q, Ma L L, Cai C L. Production of 5-hydroxymethylfurfural and furfural from lignocellulosic biomass in water-tetrahydrofuran media with sodium bisulfate. Chinese Journal of Chemical Physics, 2015, 28 (5): 650-656.

[12] Amarasekara A S, Gutierrez Reyes C D. Brønsted acidic ionic liquid catalyzed one-pot conversion of cellulose to furanic biocrude and identification of the products using LC-MS. Renewable Energy, 2019, 136: 352-357.

[13] Rasmussen H, Sørensen H R, Tanner D, Meyer A S. New pentose dimers with bicyclic moieties from pretreated biomass. Rsc Advances, 2017, 7 (9): 5206-5213.

[14] Rasmussen H, Tanner D, Sørensen H R, Meyer A S. New degradation compounds from lignocellulosic biomass pretreatment: routes for formation of potent oligophenolic enzyme inhibitors. Green Chemistry, 2017, 19: 464-473.

[15] Rasmussen H, Sorensen H R, Meyer A S. Formation of degradation compounds from lignocellulosic biomass in the biorefinery: sugar reaction mechanisms. Carbohydr Res, 2014, 385: 45-57.

[16] Rasmussen H, Mogensen K H, Jeppesen M D, Sørensen H R, Meyer A S. 4-Hydroxybenzoic acid from hydrothermal pretreatment of oil palm empty fruit bunches-its origin and influence on biomass conversion. Biomass and Bioenergy, 2016, 93: 209-216.

[17] Aguera A, Perez Estrada L A, Ferrer I, Thurman E M, Malato S, Fernandez-Alba A R. Application of time-of-flight mass spectrometry to the analysis of phototransformation products of diclofenac in water under natural sunlight. Journal of Mass Spectrometry, 2005, 40 (7): 908-915.

[18] Wang Y L, Deng W P, Wang B J, Zhang Q H, Wan X Y, Tang Z C, Wang Y, Zhu C, Cao Z X, Wang G C, Wan H L. Chemical synthesis of lactic acid from cellulose catalysed by lead (II) ions in water. Nature Communications, 2013, 4: 1-7.

[19] Luijkx G C A, Rantwijk F V, Bekkum H V. Formation of 1, 2, 4-benzenetriol by hydrothermal treatment of carbohydrates. Recueil Des Travaux Chimiques Des Pays-Bas, 1991, 110: 343-344.

[20] Girisuta B, Janssen L P B M, Heeres H J. A kinetic study on the decomposition of 5-hydroxymethylfurfural into levulinic acid. Green Chemistry, 2006, 8 (8): 701-709.

[21] Tolborg S, Meier S, Sadaba I, Elliot S G, Kristensen S K, Saravanamurugan S, Riisager A, Fristrup P, Skrydstrup T, Taarning E. Tin-containing silicates: identification of a glycolytic riathway via 3-deoxyglucosone. Green Chemistry, 2016, 18 (11): 3360-3369.

[22] Shi N, Liu Q Y, Wang T J, Ma L L, Zhang Q, Zhang Q. One-pot degradation of cellulose into furfural compounds in hot compressed steam with dihydric phosphates. Acs Sustainable Chemistry and Engineering, 2014, 2 (4): 637-642.

[23] Vom Eyser C, Schmidt T C, Tuerk J. Fate and behaviour of diclofenac during hydrothermal carbonization. Chemosphere, 2016, 153: 280-286.

[24] Kang S M, Fu J X, Zhang G. From lignocellulosic biomass to levulinic acid: a review on acid-catalyzed hydrolysis. Renewable and Sustainable Energy Reviews, 2018, 94 340-362.

[25] Heeres H J, Girisuta B, Janssen L P B M. A kinetic study on the conversion of glucose to levulinic acid. Chemical Engineering Research and Design, 2006, 84 (A5): 339-349.

[26] Horvat J, Klaic B, Metelko B, Sunjic V. Mechanism of levulinic acid formation. Tetrahedron Letters, 1985, 26 (17): 2111-2114.

[27] Shi N, Liu Q, Liu Y, Chen L, Chen N, Peng J, Ma L. Formation of soluble furanic and carbocyclic oxy-organics during the hydrothermal carbonization of glucose. Energy and Fuels, 2020, 34 (2): 1830-1840.

[28] Patil S K R, Lund C R F. Formation and growth of humins via aldol addition and condensation during acid-catalyzed conversion of 5-hydroxymethylfurfural. Energy and Fuels, 2011, 25 (10): 4745-4755.

[29] Patil S K R, Heltzel J, Lund C R F. Comparison of structural features of humins formed catalytically from glucose, fructose, and 5-hydroxymethylfurfuraldehyde. Energy and Fuels, 2012, 26 (8): 5281-5293.

[30] Dusselier M, Sels B F. Selective catalysis for cellulose conversion to lactic acid and other α-hydroxy acids. Selective Catalysis for Renewable Feedstocks and Chemicals, 2014, 353: 85-125.

[31] Holm M S, Saravanamurugan S, Taarning E. Conversion of sugars to lactic acid derivatives using heterogeneous zeotype catalysts. Science, 2010, 328 (5978): 602-605.

[32] Brands C M J, Boekel M a J S V. Reactions of monosaccharides during heating of sugar-casein systems: building of a reaction network model. Journal of Agricultural and Food Chemistry, 2001, 49 (10): 4667-4675.

[33] Tný O N, Cejpek K, Velíšek J. Formation of Carboxylic Acids during Degradation of Monosaccharides. Czech Journal of Food Sciences, 2008, 26 (2): 117-131.

[34] Ginz M, Balzer H H, Bradbury A G W, Maier H G. Formation of aliphatic acids by carbohydrate degradation during roasting of coffee. European Food Research and Technology, 2000, 211: 404-410.

[35] Li X, Xu R, Yang J, Nie S, Liu D, Liu Y, Si C. Production of 5-hydroxymethylfurfural and levulinic acid from lignocellulosic biomass and catalytic upgradation. Industrial Crops and Products, 2019, 130: 184-197.

[36] Pearson R G, Mayerle E A. Mechanism of the hydrolytic cleavage of carbon-carbon bonds. I. alkaline hydrolysis of β-diketones. Journal of the American Chemical Society, 1951, 73: 926-930.

[37] Rakete S, Berger R, Bohme S, Glomb M A. Oxidation of isohumulones induces the formation of carboxylic acids by hydrolytic cleavage. Journal of Agricultural and Food Chemistry, 2014, 62 (30): 7541-7549.

[38] Smuda M, Glomb M A. Fragmentation pathways during Maillard-induced carbohydrate degradation. Journal of Agricultural and Food Chemistry, 2013, 61 (43): 10198-10208.

[39] Smuda M, Glomb M A. Novel insights into the maillard catalyzed degradation of maltose. Journal of Agricultural and Food Chemistry, 2011, 59 (24): 13254-13264.

[40] Davidek T, Devaud S, Robert F, Blank I. Sugar fragmentation in the maillard reaction cascade: isotope labeling studies on the formation of acetic acid by a hydrolytic β-dicarbonyl cleavage mechanism. Journal of Agricultural and Food Chemistry, 2006, 54 (18): 6667-6676.

[41] Davidek T, Robert F, Devaud S, Vera F A, Blank I. Sugar fragmentation in the maillard reaction cascade: formation of short-chain carboxylic acids by a new oxidative α-dicarbonyl cleavage pathway. Journal of Agricultural and Food Chemistry, 2006, 54 (18): 6677-6684.

[42] Cao X, Peng X, Sun S, Zhong L, Sun R. Hydrothermal conversion of bamboo: identification and distribution of the components in solid residue, water-soluble and acetone-soluble fractions. Journal of Agricultural and Food Chemistry, 2014, 62 (51): 12360-12365.

[43] Kabyemela B M, Adschiri T, Malaluan R, Arai K. Degradation kinetics of dihydroxyacetone and glyceraldehyde in subcritical and supercritical water. Industrial and Engineering Chemistry Research, 1997, 36 (6): 2025-2030.

[44] Kabyemela B M, Adschiri T, Malaluan R M, Arai K. Kinetics of glucose epimerization and decomposition in subcritical and supercritical water. Industrial and Engineering Chemistry Research, 1997, 36 (5): 1552-1558.

[45] Dusselier M, van Wouwe P, de Clippel F, Dijkmans J, Gammon D W, Sels B F. Mechanistic insight into the conversion of tetrose sugars to novel α-hydroxy acid platform molecules. ChemCatChem, 2012, 5 (2): 569-575.

[46] Knill C J, Kennedy J F. Degradation of cellulose under alkaline conditions. Carbohydrate Polymers, 2003, 51 (3): 281-300.

[47] Chen H S, Wang A, Sorek H, Lewis J D, Roman-Leshkov Y, Bell A T. Production of hydroxyl-rich acids from xylose and glucose using Sn-BEA Zeolite. Chemistryselect, 2016, 1 (14): 4167-4172.

[48] Feather M S, Harris J F. Dehydration Reactions of Carbohydrates. Advances in Carbohydrate Chemistry and Biochemistry Series, 1973, 28: 161-224.

[49] Hellwig M, Degen J, Henle T. 3-deoxygalactosone, a "new" 1, 2-dicarbonyl compound in milk products. Journal of Agricultural and Food Chemistry, 2010, 58 (19): 10752-10760.

[50] Jadhav H, Pedersen C M, Solling T, Bols M. 3-Deoxy-glucosone is an intermediate in the formation of furfurals from D-glucose. ChemSusChem, 2011, 4 (8): 1049-1051.

[51] Lin H, Strull J, Liu Y, Karmiol Z, Plank K, Miller G, Guoc Z, Yang L. High yield production of levulinic acid by catalytic partial oxidation of cellulose in aqueous media. Energy and Environmental Science, 2012, 5 9773-9777.

[52] Akien G R, Qi L, Horvath I T. Molecular mapping of the acid catalysed dehydration of fructose. Chemical Communications, 2012, 48 (47): 5850-5852.

[53] Clercq R D, Dusselier M, Christiaens C, Dijkmans J, Iacobescu R I, Pontikes Y, Sels B F. Confinement effects in Lewis acid-catalyzed sugar conversion: steering toward functional poly-

ester building blocks. Acs Catalysis, 2015, 5 (10): 5803-5811.

[54] Sølvhøj A, Taarning E, Madsen R. Methyl vinyl glycolate as a diverse platform molecule. Green Chemistry. 2016, 18 (20): 5448-5455.

[55] Deng W P, Zhang Q H, Wang Y. Catalytic transformations of cellulose and cellulose-derived carbohydrates into organic acids. Catalysis Today, 2014, 234 31-41.

[56] Deng W P, Zhang Q H, Wang Y. Catalytic transformations of cellulose and its derived carbohydrates into 5-hydroxymethylfurfural, levulinic acid, and lactic acid. Science China—Chemistry, 2015, 58 (1): 29-46.

附 录

葡萄糖及生物质衍生呋喃类物质水热转化生成的可溶性物质的MS和MS^{2}数据及谱图

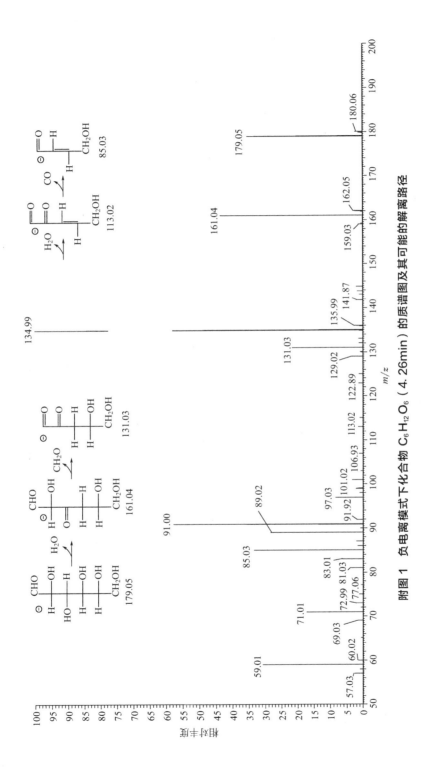

附图 1 负电离模式下化合物 $C_6H_{12}O_6$（4.26min）的质谱图及其可能的解离路径

附录 葡萄糖及生物质衍生呋喃类物质水热转化生成的可溶性物质的 MS 和 MS^2 数据及谱图

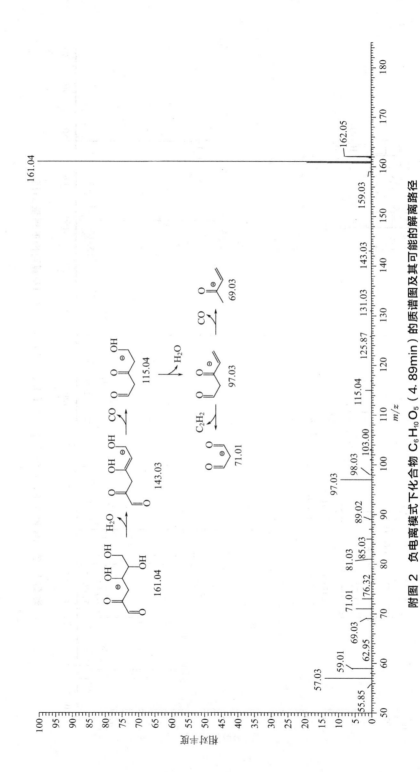

附图 2 负电离模式下化合物 $C_6H_{10}O_5$（4.89min）的质谱图及其可能的解离路径

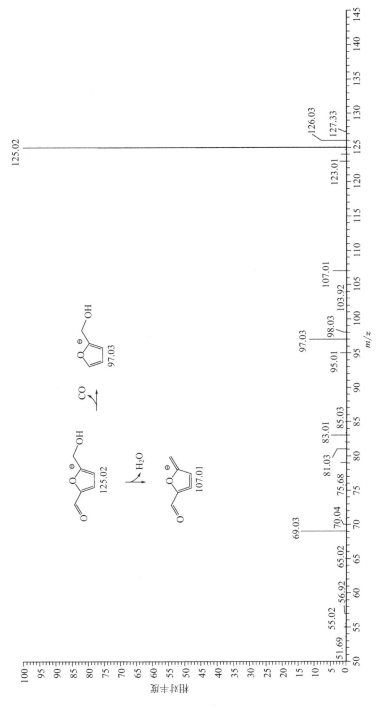

附图 3 负电离模式下化合物 $C_6H_6O_3$（10.17min）的质谱图及其可能的解离路径

附录 葡萄糖及生物质衍生呋喃类物质水热转化生成的可溶性物质的 MS 和 MS^2 数据及谱图

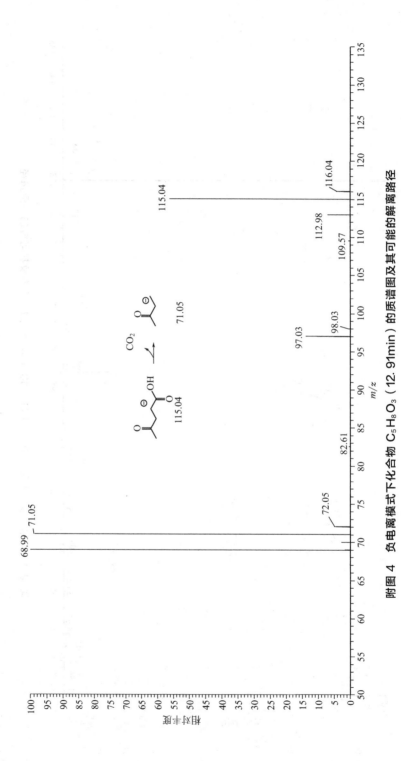

附图 4 负电离模式下化合物 $C_5H_8O_3$（12.91min）的质谱图及其可能的解离路径

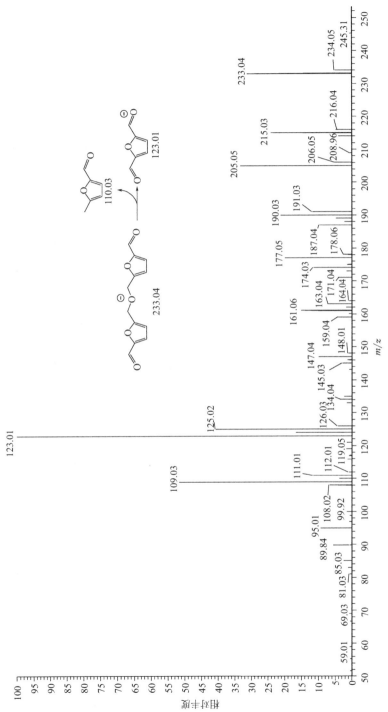

附图 5　负电离模式下化合物 $C_{12}H_{10}O_5$（15.45min）的质谱图及其可能的解离路径

附录　葡萄糖及生物质衍生呋喃类物质水热转化生成的可溶性物质的 MS 和 MS^2 数据及谱图

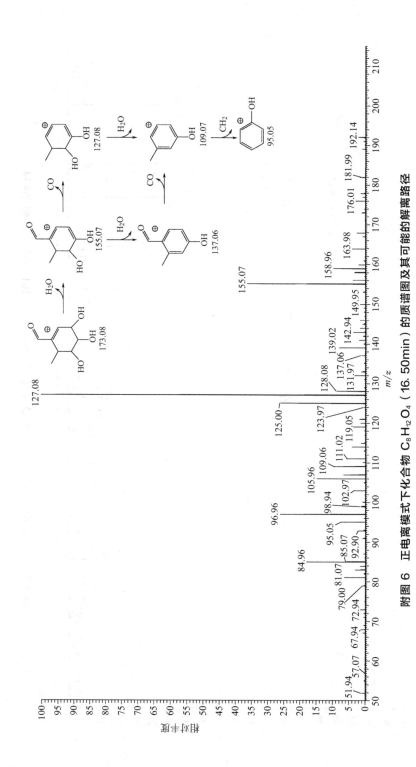

附图6 正电离模式下化合物 $C_8H_{12}O_4$（16.50min）的质谱图及其可能的解离路径

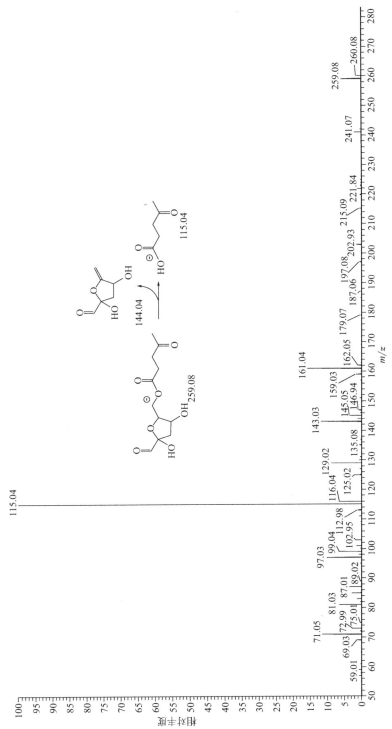

附图 7 负电离模式下化合物 $C_{11}H_{16}O_7$ (17.97min) 的质谱图及其可能的解离路径

附录 葡萄糖及生物质衍生呋喃类物质水热转化生成的可溶性物质的 MS 和 MS^2 数据及谱图

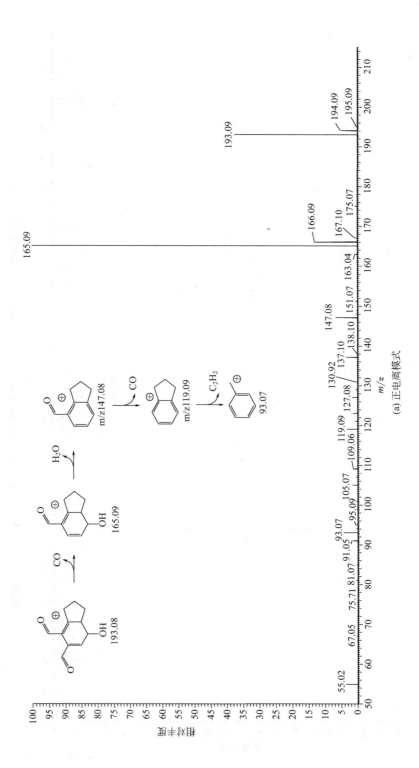

(a) 正电离模式

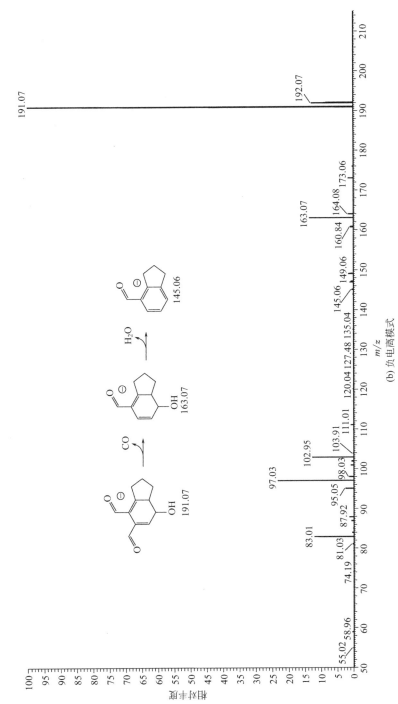

附图8 化合物 $C_{11}H_{12}O_3$ (18.11min) 的质谱图及其可能的解离路径

附录 葡萄糖及生物质衍生呋喃类物质水热转化生成的可溶性物质的 MS 和 MS^2 数据及谱图

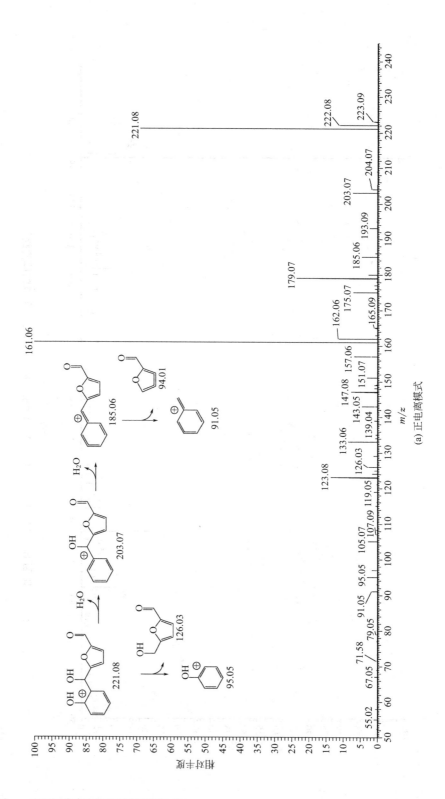

(a) 正电离模式

190　木质纤维素水热炼制原理与技术

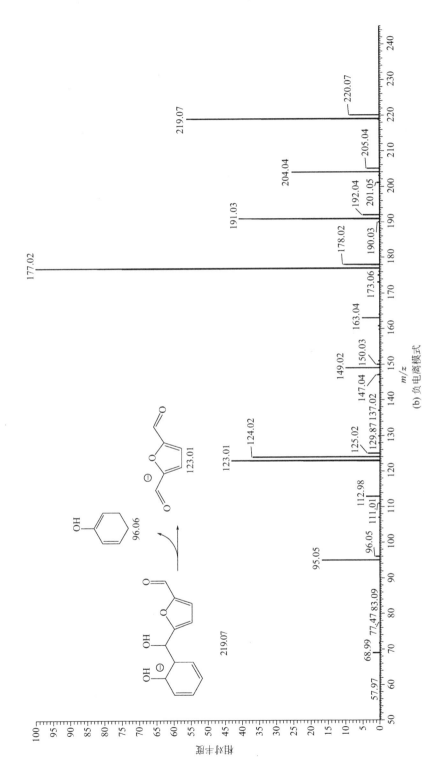

附图 9 化合物 $C_{12}H_{12}O_4$ (19.81min) 的质谱图及其可能的解离路径

附录 葡萄糖及生物质衍生呋喃类物质水热转化生成的可溶性物质的 MS 和 MS^2 数据及谱图

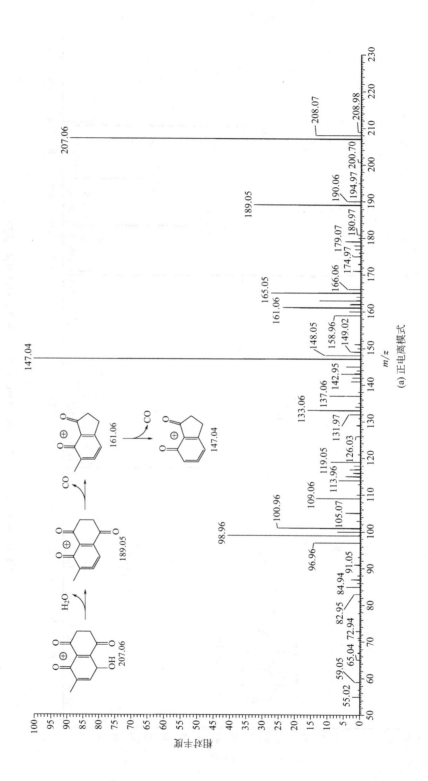

(a) 正电离模式

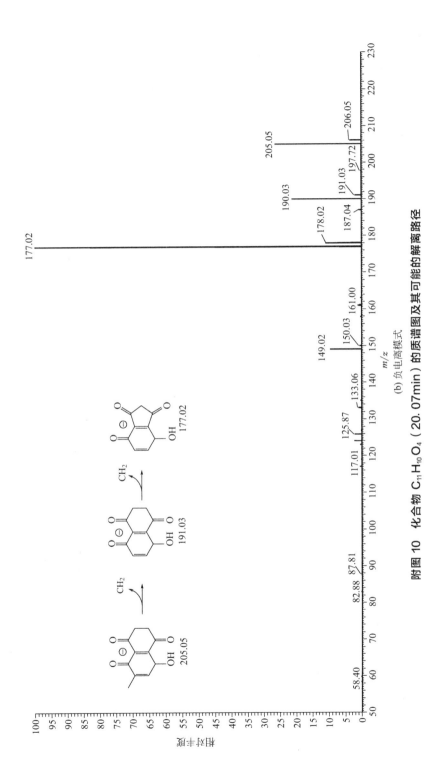

附图 10 化合物 $C_{11}H_{10}O_4$（20.07min）的质谱图及其可能的解离路径

附录 葡萄糖及生物质衍生呋喃类物质水热转化生成的可溶性物质的 MS 和 MS^2 数据及谱图

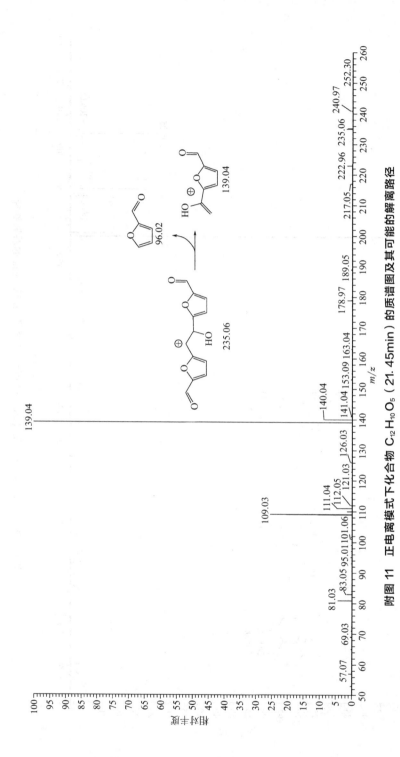

附图11 正电离模式下化合物 $C_{12}H_{10}O_5$（21.45min）的质谱图及其可能的解离路径

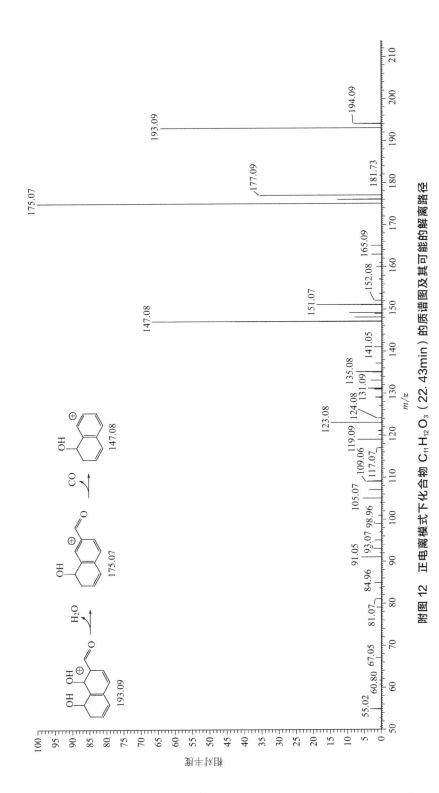

附图 12　正电离模式下化合物 $C_{11}H_{12}O_3$（22.43min）的质谱图及其可能的解离路径

附录　葡萄糖及生物质衍生呋喃类物质水热转化生成的可溶性物质的 MS 和 MS^2 数据及谱图

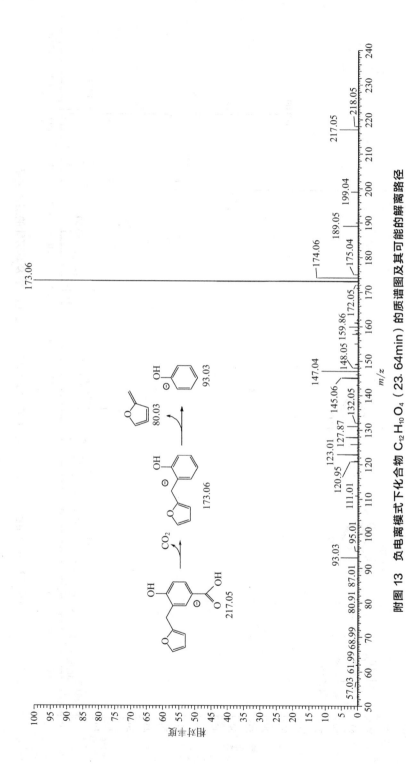

附图13 负电离模式下化合物 $C_{12}H_{10}O_4$（23.64min）的质谱图及其可能的解离路径

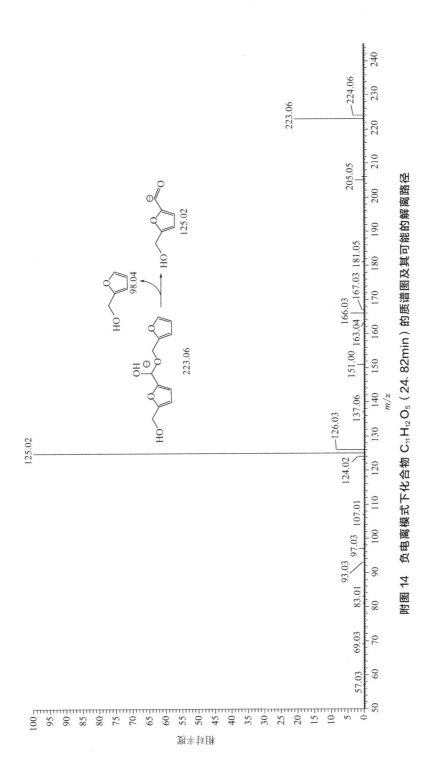

附图 14　负电离模式下化合物 $C_{11}H_{12}O_5$（24.82min）的质谱图及其可能的解离路径

附录　葡萄糖及生物质衍生呋喃类物质水热转化生成的可溶性物质的 MS 和 MS^2 数据及谱图

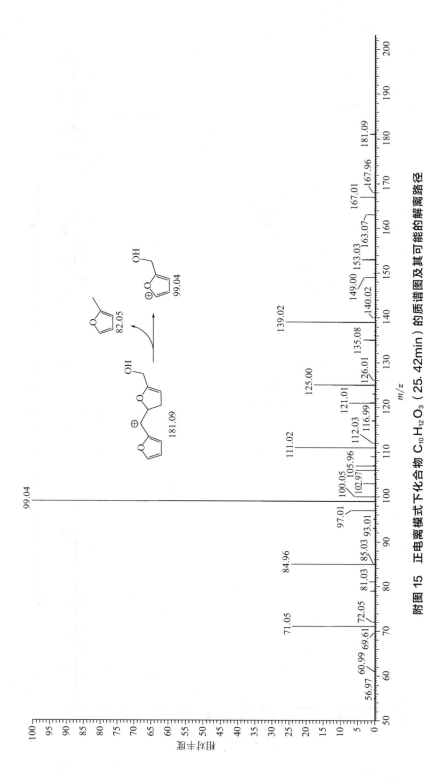

附图15 正电离模式下化合物 $C_{10}H_{12}O_3$（25.42min）的质谱图及其可能的解离路径

198　木质纤维素水热炼制原理与技术

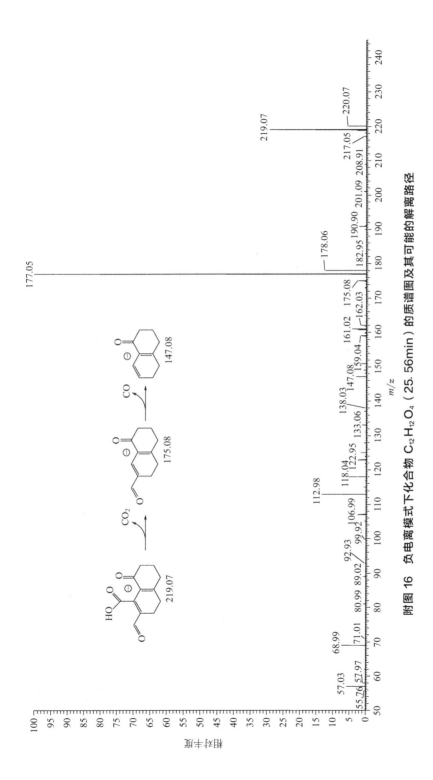

附图 16 负电离模式下化合物 $C_{12}H_{12}O_4$（25.56min）的质谱图及其可能的解离路径

附录 葡萄糖及生物质衍生呋喃类物质水热转化生成的可溶性物质的 MS 和 MS^2 数据及谱图

附表 1　HMF 水热转化过程中生成的可溶性物质的 MS 和 MS² 数据

停留时间/min	分子式	分子量	电离模式	MS	MS²
10.20	$C_6H_6O_3$	126.03	负电离模式	125.02	125.02,123.01,107.01,97.03,95.01,83.01,79.02,70.04,69.03
12.98	$C_5H_8O_3$	116.05	正电离模式	117.05	117.05,99.04,81.03,71.05
15.48	$C_{12}H_{10}O_5$	234.05	负电离模式	233.04	233.04,205.05,191.03,187.04,177.05,174.03,171.04,163.04,161.06,159.04,149.02,147.04,145.03,134.04,125.02,123.01,121.03,111.01,109.03,97.03,95.01,85.03
18.07~18.13	$C_{11}H_{12}O_3$	192.08	正电离模式	193.09	193.09,175.08,165.09,163.04,147.08,138.1,137.1,127.08,119.09,105.07,99.04,93.07,91.05
19.93~19.96	$C_{12}H_{12}O_4$	220.07	负电离模式	191.07	191.07,173.06,163.08,149.06,135.04,121.06,111.01,97.03,95.05,87.01,83.01,81.03
			正电离模式	221.08	221.08,203.07,193.09,179.07,175.08,161.06,151.07,147.08,133.06,123.08,105.07,95.05
			负电离模式	219.07	—
21.48~21.50	$C_{12}H_{10}O_5$	234.05	正电离模式	235.06	235.06,189.06,171.04,139.04,126.03,111.04,109.03,97.03,81.03
			负电离模式	233.04	233.04,187.04,175.04,161.02,159.04,149.02,137.02,135.09,124.02,109.03,95.01,71.01
22.56	$C_{11}H_{14}O_3$	194.09	正电离模式	195.10	195.10,177.09,159.08,149.1,141.05,135.08,131.09,121.1,107.09,105.07,95.05,93.07,91.05,81.07,79.05
23.68	$C_{12}H_{10}O_4$	218.06	负电离模式	217.05	217.05,189.05,173.06,161.06,131.09,125.1,121.1,110.07,109.07,107.09,105.07,97.1,93.07
24.16	$C_{10}H_{14}O_2$	166.10	正电离模式	167.11	167.11,149.1,137.06,131.09,125.1,121.1,1110.07,109.07,105.07,97.07,93.07,91.05,79.05
24.75~24.85	$C_{11}H_{12}O_5$	224.07	正电离模式	225.08	225.08,207.07,189.05,179.03,161.06,153.05,135.04,121.06,109.07,99.04,93.07,81.03,71.05
			负电离模式	223.06	223.06,205.05,187.04,181.05,163.04,151,137.06,126.03,125.02,124.02,107.01,97.03,81.03,69.03

续表

停留时间/min	分子式	分子量	电离模式	MS	MS²
25.49~25.60	$C_{10}H_{12}O_3$	220.07	正电离模式	221.08	219.07,217.05,201.06,175.08,163.03,161.02,149.02,133.03,89.02
			负电离模式	219.07	—
26.37	$C_{12}H_{12}O_3$	204.08	正电离模式	205.09	205.09,191.07,189.05,187.08,163.04,151.08,139.04,123.08,109.07,99.04,95.05,85.07,71.05

附表 2 糠醛水热转化过程中生成的可溶性物质的 MS 和 MS² 数据

停留时间/min	分子式	分子量	电离模式	MS	MS²
10.13	$C_5H_8O_4$	132.04	负电离模式	131.03	131.03,129.02,113.02,85.03
10.57	$C_4H_6O_4$	118.02	负电离模式	117.02	117.02,99.01,73.03
16.60	$C_6H_8O_2$	112.05	负电离模式	111.04	111.04,93.03,83.05
17.01	$C_{10}H_8O_4$	192.04	负电离模式	191.03	191.03,173.03,163.04,147.04,145.03,117.03
18.39	$C_{12}H_{14}O_4$	222.09	负电离模式	221.08	221.08,203.07,189.02,179.03,161.02,145.03,137.06,133.03,123.04,117.03,105.03,97.03,95.05
19.12	$C_9H_{12}O_3$	168.07	负电离模式	167.07	167.07,153.05,152.05,149.06,126.06,123.04,121.06,110.04,97.03,83.01
19.80	$C_9H_{12}O_4$	184.07	负电离模式	183.06	183.06,153.05,137.02,119.01,109.03,93.03,81.03
20.94	$C_{11}H_{12}O_3$	192.04	正电离模式	193.05	193.09,177.09,165.09,163.04,160.05,151.08,149.06,147.08,135.04,107.05,99.04,91.05,79.06
			负电离模式	191.03	191.03,175.04,165.05,163.04,151.04,149.02,147.04,135.04,123.04,119.05,117.03,107.05,95.01
21.50	$C_{15}H_{12}O_5$	272.07	正电离模式	273.08	273.08,255.07,245.08,231.07,227.09,217.09,203.07,199.08,189.09,185.06,175.08,171.08,161.1,157.07,147.08,143.09,129.07,119.09,109.03,107.05,95.05,91.05
			负电离模式	271.06	271.06,243.07,227.07,215.07,201.06,199.08,197.06,171.08,161.02,147.04,145.03,133.03,123.04,121.03,97.03,93.03

续表

停留时间/min	分子式	分子量	电离模式	MS	MS²
22.21	$C_{10}H_{10}O_3$	178.06	正电离模式	179.07	179.07,161.06,151.08,137.06,133.07,121.01,111.02,105.07
23.26	$C_{10}H_8O_4$	192.04	正电离模式	193.05	193.05,175.04,165.06,147.04,137.06,131.05,123.04,119.05,113.06,109.07,103.05, 95.05,91.06,81.03,79.06
24.13	$C_{10}H_8O_4$	192.04	负电离模式	193.05	193.05, 175.04, 165.06, 147.04, 137.06, 125.02, 123.04, 119.05, 109.07, 96.05, 95.05,91.06
24.13				191.03	191.03,165.05,163.04,147.04,135.04,133.03,121.03,119.05,107.05,106.04,93.03, 91.05,85.02
25.22	$C_{10}H_8O_4$	192.04	正电离模式	193.05	193.05,177.06,149.06,135.04,131.09,121.07,107.09,93.07,85.03,79.06
25.22				191.03	191.03,163.04,147.04,135.04,121.03,119.05,107.05,91.05,77.04
25.57	$C_{11}H_{10}O_5$	222.05	正电离模式	223.06	223.06,205.05,191.03,175.04,165.06,149.06,123.01,121.07,111.04,109.07,99.04, 95.01,93.07,91.05,79.06
26.28~26.38	$C_{15}H_{12}O_5$	272.07	正电离模式	273.08	273.08,245.08,217.06,201.06,195.07,179.07,161.06,151.08,149.06,133.07,123.08, 121.01,109.07,105.07,79.06
26.28~26.38			负电离模式	271.06	271.06,243.07,227.07,215.07,209.06,201.06,185.06,173.06,157.07,147.04, 143.05,133.03,123.04,121.07,97.03,93.03

附表 3 糖醇水热转化过程中生成的可溶性物质的 MS 和 MS² 数据

停留时间/min	分子式	分子量	电离模式	MS	MS²
10.53	$C_4H_6O_4$	118.03	负电离模式	117.02	117.02,99.01,73.03
12.99	$C_5H_8O_3$	116.05	正电离模式	117.06	117.06,99.04,71.05
13.45	$C_5H_6O_2$	98.04	负电离模式	115.04	115.04,97.03,71.05
13.45				97.03	97.03,79.02,69.03
15.46	$C_8H_{10}O_3$	154.06	正电离模式	155.07	155.07,137.06,127.08,119.05,113.02,109.07,95.05,93.07,85.07,81.07
15.46			负电离模式	153.05	153.05,135.04,125.06,111.04,109.03,93.03

续表

停留时间/min	分子式	分子量	电离模式	MS	MS^2
16.03	$C_{10}H_{12}O_4$	196.07	负电离模式	195.07	195.07,151.08,135.04,133.06,109.06,95.05,59.01
16.53	$C_8H_{10}O_3$	154.06	正电离模式	155.07	155.07,139.08,127.08,109.07,97.07,95.05,93.07,85.07,81.07,79.06
17.20	$C_{10}H_{12}O_4$	196.07	正电离模式	197.08	197.08,179.07,161.06,151.08,123.04,109.07,101.02,97.07,79.06,73.03
18.01	$C_{10}H_{12}O_4$	196.07	负电离模式	195.07	195.07,177.06,151.08,133.06,109.06,95.05
			正电离模式	197.08	197.08,179.07,161.06,151.08,133.06,119.05,109.07,105.07,93.07
18.71	$C_9H_{14}O_3$	170.09	负电离模式	195.07	195.07,177.06,151.08,149.06,133.06,123.04,109.06,95.05,93.03
			正电离模式	171.10	171.1,154.1,153.09,135.09,113.08,95.05,93.07
19.40	$C_{10}H_{12}O_4$	196.07	正电离模式	197.08	197.08,179.07,155.07,151.08,137.06,133.07,119.05,109.07,105.07,93.07,81.07
			负电离模式	195.07	195.07,151.08,149.06,135.04,133.06,109.06,107.05,95.05,93.03,91.05
19.98	$C_{10}H_{10}O_3$	178.06	正电离模式	179.07	179.07,161.06,151.08,137.06,133.07,123.04,109.07,95.05
			负电离模式	355.12	177.06,149.06,133.06,121.03,95.05
21.12	$C_{15}H_{14}O_4$	258.09	正电离模式	259.10	259.1,241.09,231.1,217.09,213.09,203.11,199.08,189.09,185.1,177.09,171.08,161.1,149.06,143.09,133.07,129.07,121.07,111.04,107.05,105.07,95.05,83.05
21.42	$C_{11}H_{12}O_3$	192.08	负电离模式	257.08	257.08,239.07,229.09,213.09,211.08,195.08,185.06,176.05,159.04,147.04,97.03
22.87	$C_{14}H_{12}O_2$	212.08	正电离模式	193.09	193.09,177.09,175.08,165.09,161.06,151.08,149.10,139.08,133.07,125.06,123.08,105.07,95.05,91.05
26.37	$C_{10}H_{10}O_3$	178.06	正电离模式	213.09	213.09,195.08,185.1,171.08,167.09,165.09,157.1,143.09,133.07,131.05,129.07,128.06,119.05,105.07,95.05,91.06,81.03
27.29	$C_{11}H_{12}O_3$	212.08	正电离模式	179.07	179.07,161.06,151.08,133.07,123.08,109.07,95.05
27.86	$C_{14}H_{16}O_3$	232.11	正电离模式	213.09	213.09,211.08,193.09,185.1,171.08,165.09,157.1,143.09,137.06,129.07,115.04,111.08,105.07,101.02,95.05
				233.12	233.12,215.11,213.09,205.12,197.1,191.11,190.1,187.11,185.1,175.08,173.1,169.1,163.11,153.07,145.1,133.07,131.09,123.08,109.07,99.04,83.05